CLIMATE CHANGE
INDICATORS
IN THE UNITED STATES
2016
FOURTH EDITION

Find Us Online

Please visit EPA's website at: www.epa.gov/climate-indicators.
There you can:

- View the latest information about EPA's climate change indicators
- View additional graphs, maps, and background information
- Access corresponding technical documentation
- Download images and figures
- Suggest new indicators for future reports

You can also send an email to: climateindicators@epa.gov.

Suggested Citation

U.S. Environmental Protection Agency. 2016.
Climate change indicators in the United States, 2016. Fourth edition.
EPA 430-R-16-004. www.epa.gov/climate-indicators.

Contents

Acknowledgments

DATA CONTRIBUTORS AND INDICATOR REVIEWERS

EPA wishes to thank the federal government agencies, nongovernmental organizations, and other institutions that participate in the ongoing EPA Indicators project. Their commitment, contributions, and collaboration have helped to make this report possible.

U.S. Governmental Organizations

- Centers for Disease Control and Prevention: C. Ben Beard, Lars Eisen, Micah Hahn, George Luber, Ambarish Vaidyanathan
- National Aeronautics and Space Administration: Joey Comiso, Stacey Frith, Thorsten Markus, Walt Meier
- National Oceanic and Atmospheric Administration
  - Climate Prediction Center: Gerry Bell
  - Earth System Research Laboratory: Ed Dlugokencky, Steve Montzka
  - National Marine Fisheries Service: Roger Griffis
  - National Centers for Environmental Information: Deke Arndt, Karin Gleason, Boyin Huang, John Marra
  - National Ocean Service: William Sweet, Chris Zervas
- U.S. Department of Agriculture
  - Agricultural Research Service: Lewis Ziska
  - Forest Service: Jennifer Lecker, Karen Short
- U.S. Geological Survey
  - Alaska Science Center: Shad O'Neel, Louis Sass
  - Maine Water Science Center: Robert Dudley, Glenn Hodgkins
  - New York Water Science Center: Michael McHale
  - Virginia Water Science Center: John Jastram, Karen Rice
  - Washington Water Science Center: Matt Bachmann

Universities, Nongovernmental Organizations, and International Institutions

- Bermuda Institute of Ocean Sciences: Nick Bates
- Commonwealth Scientific and Industrial Research Organisation: John Church, Catia Domingues, Didier Monselesan, Neil White
- Georgia Institute of Technology: Ray Wang
- Japan Agency for Marine-Earth Science and Technology: Masayoshi Ishii
- Massachusetts Institute of Technology: Kerry Emanuel
- National Audubon Society: Justin Schuetz, Candan Soykan
- National Physical Laboratory/University of Edinburgh: Tim Arnold
- North Carolina State University: Ken Kunkel
- Oregon State University, The Oregon Climate Change Research Institute: Philip Mote, Darrin Sharp
- Rutgers University, Global Snow Lab: David Robinson; Marine and Coastal Sciences: Malin Pinsky
- Scripps Institution of Oceanography, University of California San Diego: Jens Mühle
- Universidad de las Palmas de Gran Canaria: Melchor González-Dávila
- University of Bristol: Matthew Rigby
- University of Colorado: Mark Tschudi
- University of Iowa: Iman Mallakpour, Louise Slater, Gabriele Villarini
- University of Montana: John Dore
- University of Nebraska-Lincoln: Song Feng
- University of Wisconsin-Madison: Corinna Gries
- University of Wisconsin-Milwaukee: Mark Schwartz
- Woods Hole Oceanographic Institution: Ivan Lima
- World Glacier Monitoring Service: Michael Zemp
- World Resources Institute: Tom Damassa

PEER REVIEW

The report was peer reviewed by 11 external, independent experts: Christopher M. Barker, Kristie L. Ebi, Andrew J. Elmore, Anthony Janetos, Paul Kirshen, Noah Molotch, Bart Ostro, Charles H. Peterson, Kenneth W. Potter, Mark Serreze, and Tanja Srebotnjak.

REPORT DEVELOPMENT AND PRODUCTION

Overall coordination and development of this report was provided by EPA's Office of Atmospheric Programs, Climate Change Division. Support for content development, data analysis, and report design and production was provided by Eastern Research Group, Inc. (ERG). Abt Associates also provided analytical and content development support.

Introduction

The Earth's climate is changing. Temperatures are rising, snow and rainfall patterns are shifting, and more extreme climate events—like heavy rainstorms and record-high temperatures—are already taking place. Scientists are highly confident that many of these observed changes can be linked to the levels of carbon dioxide and other greenhouse gases in our atmosphere, which have increased because of human activities.

HOW IS THE CLIMATE CHANGING?

Since the Industrial Revolution began in the 1700s, people have added a significant amount of greenhouse gases into the atmosphere, largely by burning fossil fuels to generate electricity, heat and cool buildings, and power vehicles—as well as by clearing forests. The major greenhouse gases that people have added to the atmosphere are carbon dioxide, methane, nitrous oxide, and fluorinated gases. When these gases are emitted into the atmosphere, many remain there for long time periods, ranging from a decade to thousands of years. Past emissions affect our atmosphere in the present day; current and future emissions will continue to increase the levels of these gases in our atmosphere for the foreseeable future.

"Greenhouse gases" got their name because they trap heat (energy) like a greenhouse in the lower part of the atmosphere (see "The Greenhouse Effect" below). As more of these gases are added to the atmosphere, more heat is trapped. This extra heat leads to higher air temperatures near the Earth's surface, alters weather patterns, and raises the temperature of the oceans.

These observed changes affect people and the environment in important ways. For example, sea levels are rising, glaciers are melting, and plant and animal life cycles are changing. These types of changes can bring about fundamental disruptions in ecosystems, affecting plant and animal populations, communities, and biodiversity. Such changes can also affect people's health and quality of life, including where people can live, what kinds of crops are most viable, what kinds of businesses can thrive in certain areas, and the condition of buildings and infrastructure. Some of these changes may be beneficial to certain people and places, as indicators like **Length of Growing Season** point out. Over time, though, many more of these changes will have negative consequences for people and society.[1]

What Is Climate Change?

Climate change refers to any substantial change in measures of climate (such as temperature or precipitation) lasting for an extended period (decades or longer). Natural factors have caused the climate to change during previous periods of the Earth's history, but human activities are the primary cause of the changes that are being observed now.

Global warming is a term often used interchangeably with the term "climate change," but they are not entirely the same thing. Global warming refers to an average increase in the temperature of the atmosphere near the Earth's surface. Global warming is just one aspect of global climate change, though a very important one.

Why Use Indicators?

One important way to track and communicate the causes and effects of climate change is through the use of indicators. An indicator represents the state or trend of certain environmental or societal conditions over a given area and a specified period of time. For example, long-term measurements of temperature in the United States and globally are used as an indicator to track and better understand the effects of changes in the Earth's climate.

How Do the Indicators Relate to Climate Change?

All of the indicators in this report relate to either the causes or effects of climate change. Some indicators show trends that can be more directly linked to human-induced climate change than others. Collectively, the trends depicted in these indicators provide important evidence of "what climate change looks like."

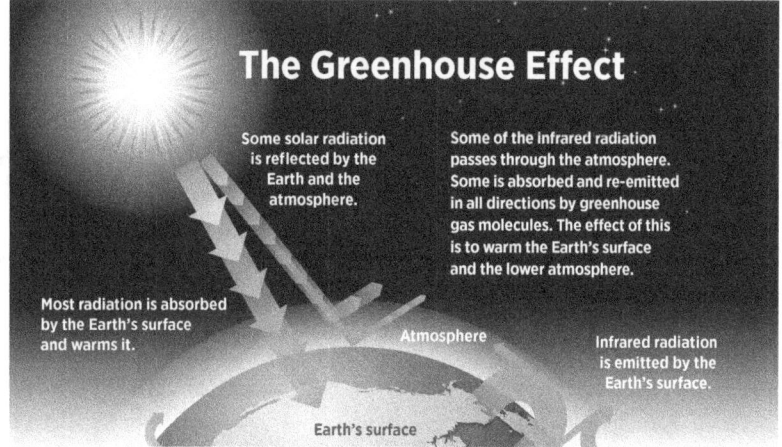

The Greenhouse Effect

Some solar radiation is reflected by the Earth and the atmosphere.

Some of the infrared radiation passes through the atmosphere. Some is absorbed and re-emitted in all directions by greenhouse gas molecules. The effect of this is to warm the Earth's surface and the lower atmosphere.

Most radiation is absorbed by the Earth's surface and warms it.

Atmosphere

Infrared radiation is emitted by the Earth's surface.

Earth's surface

ABOUT THIS REPORT

The U.S. Environmental Protection Agency (EPA) publishes this report to communicate information about the science and impacts of climate change, assess trends in environmental quality, and inform decision-making. *Climate Change Indicators in the United States, 2016,* is the fourth edition of a report first published by EPA in 2010. This report presents 37 indicators to help readers understand changes observed from long-term records related to the causes and effects of climate change, the significance of these changes, and their possible consequences for people, the environment, and society. While the indicators presented in this report do not cover all possible measures of the causes and effects of climate change, as might be found in the full body of scientific literature, they represent a wide-ranging set of indicators that show observed changes in the Earth's climate system and several climate-relevant impacts.

About EPA's Indicators

Each of EPA's 37 indicators covers a specific climate-related topic, such as **U.S. Greenhouse Gas Emissions**. Some indicators present a single measure or variable; others have multiple measures, reflecting different data sources or different ways to group, characterize, or zoom in on the data. EPA follows an established framework to identify data sets, select indicators, obtain independent expert review, and publish this report.

Data sources: All of EPA's indicators are based on peer-reviewed, publicly available data from government agencies, academic institutions, and other organizations. In addition to being published here, these data sets have been published in the scientific literature and in other government or academic reports. EPA also received input from scientists, researchers, and communications experts in nongovernmental and private sectors during the compilation of this report.

Indicator selection: EPA carefully screened and selected each indicator using a standard set of criteria that consider usefulness, data quality, and relevance to climate change. This process ensures that all indicators are based on credible data. For more information about EPA's indicator criteria and selection process, see the technical support document available at: www.epa.gov/climate-indicators.

Expert review: This report, along with all of EPA's climate change indicators and supporting documentation, was peer-reviewed by independent technical experts.

Publication: This report, the corresponding website, and the accompanying detailed technical documentation have been designed to ensure that the indicators are presented and documented clearly and transparently.

All of EPA's climate change indicators relate to either the causes or effects of climate change. Some indicators are more directly influenced by climate than others (e.g., indicators related to health outcomes), yet they all have met EPA's criteria and have a scientifically based relationship to climate. This report does not attempt to identify the *extent* to which climate change is causing a trend in an observed indicator. Connections between human activities, climate change, and observed indicators are explored in more detail elsewhere in the scientific literature.

EPA's indicators generally cover broad geographic scales and many years of data, as this is the most appropriate way to view trends relevant to climate change. After all, the Earth is a complex system, and there will always be natural variations from one year to the next—for example, a very warm year followed by a colder year. The Earth's climate also goes through other natural cycles that can play out over a period of several years or even decades. Thus, EPA's indicators present trends for as many years as the underlying data allow.

For more information, see "Frequently Asked Questions About EPA's Climate Change Indicators," available at: www.epa.gov/climate-indicators/frequent-questions.

How Is This Report Useful?

Climate Change Indicators in the United States, 2016, is written with the primary goal of informing readers' understanding of climate change. It is also designed to be useful for the public, scientists, analysts, decision-makers, educators, and others who can use climate change indicators as a tool for:

- Effectively communicating relevant climate science information in a sound, transparent, and easy-to-understand way.
- Assessing trends in environmental quality, factors that influence the environment, and effects on ecosystems and society.
- Informing science-based decision-making.

A Roadmap to the Report

The indicators are grouped into six chapters: Greenhouse Gases, Weather and Climate, Oceans, Snow and Ice, Health and Society, and Ecosystems. Some chapters also include a "Community Connection," "Tribal Connection," or "A Closer Look" feature that highlights a specific region, data record, or area of interest. Several indicators highlight the important ways in which the observed changes can have implications for human health.

Each indicator in this report fills one or two pages, and contains:

• One or more graphics depicting changes over time.
• Background on how the indicator relates to climate change.
• What's Happening: Key points about what the indicator shows.
• About the Indicator: A description of the data source and how the indicator was developed.

Additional resources that can provide readers with more information appear at the end of the report (see Climate Change Resources on p. 84).

EPA's website provides a more complete version of each indicator, including more background information, additional graphs or maps in some cases, downloadable data, interactive maps and animations for selected indicators, and important notes to help readers interpret the data. EPA also compiles an accompanying **technical support document** containing more detailed information about each indicator, including data sources, data collection methods, calculations, statistical considerations, and sources of uncertainty. This document also describes EPA's approach and criteria for selecting indicators for the report. All of this information is available on EPA's website at: www.epa.gov/climate-indicators.

WHAT'S NEW IN 2016?

The 2016 report reflects previously reported indicators and has added the following new indicators and features:

• **Seven new indicators: River Flooding, Coastal Flooding, Antarctic Sea Ice, Heat-Related Illnesses, West Nile Virus, Stream Temperature, and Marine Species Distribution.**
• **Three expanded indicators: Arctic Sea Ice** was expanded to look at changes in the length of the melt season. Similarly, **Snow Cover** now examines changes in the length of the snow cover season. **Heat-Related Deaths** has a new graph that focuses on heat-related cardiovascular disease deaths, including trends for specific at-risk groups.
• **Updated indicators:** Nearly all indicators have been updated with additional years of data that have become available since the last report.
• **Tribal connection:** The report includes an example of stream temperature trends in the Pacific Northwest and highlights how changes may affect salmon, a tribally important resource.

LOOKING AHEAD

Indicators of climate change are expected to become even more numerous and to depict even clearer trends in the future. EPA will continue to work in partnership with coordinating bodies, such as the U.S. Global Change Research Program, and with other agencies, organizations, and individuals to collect and communicate useful data and to inform policies and programs based on this knowledge. As new and more comprehensive indicator data become available, EPA will continually update the indicators presented in this report.

Understanding the Connections Between Climate Change and Human Health

It can be tempting to think of climate change as something that affects other places, other people, or something in the distant future. However, climate change already poses a very real threat to the American people. One of the biggest concerns is its effect on human health.

Scientists' understanding of how climate change increases risks to human health has advanced significantly in recent years. In April 2016, the U.S. Global Change Research Program (USGCRP) published the largest-ever assessment of the state of the science.[2] Its conclusion: Every American is vulnerable to the health impacts associated with climate change.

As the impacts increase, and as we learn more about them and how best to track them over time, government agencies and communities are also finding new ways to respond to climate-related threats. In recognition of the growing body of evidence about the health risks of climate change, this edition of EPA's climate change indicators report includes new indicators on several health-related topics, along with a special section on the connections between climate change and health (beginning on p. 53). In addition, several indicators include boxes that highlight topics related to human health. These topics are highlighted with the following icon:

Note, however, that improved understanding of human health risks due to climate change does not necessarily correspond to increased long-term data for trend assessment. The USGCRP's Climate and Health Assessment identifies the importance of long-term environmental health data and monitoring.

Understanding Greenhouse Gases

MAJOR GREENHOUSE GASES ASSOCIATED WITH HUMAN ACTIVITIES

The major greenhouse gases emitted into the atmosphere are carbon dioxide, methane, nitrous oxide, and fluorinated gases (see the table below). Some of these gases are produced almost entirely by human activities; others come from a combination of natural sources and human activities.

Many of the major greenhouse gases can remain in the atmosphere for tens to thousands of years after being released. They become globally mixed in the lower part of the atmosphere, called the troposphere (the first several miles above the Earth's surface), reflecting the combined contributions of emissions sources worldwide from the past and present. Due to this global mixing, the impact of emissions of these gases does not depend on where in the world they are emitted. Also, concentrations of these gases are similar regardless of where they are measured, as long as the measurement is far from any large sources or sinks of that gas.

Some other substances have much shorter atmospheric lifetimes (i.e., less than a year) but are still relevant to climate change. Important short-lived substances that affect the climate include water vapor, ozone in the troposphere, pollutants that lead to ozone formation, and aerosols (atmospheric particles) such as black carbon and sulfates. Water vapor, tropospheric ozone, and black carbon contribute to warming, while other aerosols produce a cooling effect. Because these substances are short-lived, their climate impact can be influenced by the location of their emissions, with concentrations varying greatly from place to place.

Several factors determine how strongly a particular greenhouse gas affects the Earth's climate. One factor is the length of time that the gas remains in the atmosphere. A second factor is each gas's unique ability to absorb energy. By considering both of these factors, scientists calculate a gas's global warming potential, which measures how much a given amount of the greenhouse gas is estimated to contribute to global warming over a specific period of time (for example, 100 years) after being emitted. For purposes of comparison, global warming potential values are calculated in relation to carbon dioxide, which is assigned a global warming potential equal to 1. The table below describes sources, lifetimes, and global warming potentials for several important long-lived greenhouse gases.

Gases and Substances Included in This Report

This report focuses on most of the major, well-mixed greenhouse gases that contribute to the vast majority of warming of the climate. It also includes certain substances with shorter atmospheric lifetimes (i.e., less than a year) that are relevant to climate change. In addition to several long-lived greenhouse gases, the online version of the **Atmospheric Concentrations of Greenhouse Gases** indicator tracks concentrations of ozone in the layers of the Earth's atmosphere, while Figure 2 of the **Climate Forcing** indicator on EPA's website shows the influence of a variety of short-lived substances.

Major Long-Lived Greenhouse Gases and Their Characteristics

Greenhouse gas	How it's produced	Average lifetime in the atmosphere	100-year global warming potential
Carbon dioxide	Emitted primarily through the burning of fossil fuels (oil, natural gas, and coal), solid waste, and trees and wood products. Changes in land use also play a role. Deforestation and soil degradation add carbon dioxide to the atmosphere, while forest regrowth takes it out of the atmosphere.	see below*	1
Methane	Emitted during the production and transport of oil and natural gas as well as coal. Methane emissions also result from livestock and agricultural practices and from the anaerobic decay of organic waste in municipal solid waste landfills.	12.4 years	28–36
Nitrous oxide	Emitted during agricultural and industrial activities, as well as during combustion of fossil fuels and solid waste.	121 years	265–298
Fluorinated gases	A group of gases that contain fluorine, including hydrofluorocarbons, perfluorocarbons, and sulfur hexafluoride, among other chemicals. These gases are emitted from a variety of industrial processes and commercial and household uses and do not occur naturally. Sometimes used as substitutes for ozone-depleting substances such as chlorofluorocarbons (CFCs).	A few weeks to thousands of years	Varies (the highest is sulfur hexafluoride at 23,500)

This table shows 100-year global warming potentials, which describe the effects that occur over a period of 100 years after a particular mass of a gas is emitted. Global warming potentials and lifetimes come from the Intergovernmental Panel on Climate Change's Fifth Assessment Report.[1]

* Carbon dioxide's lifetime cannot be represented with a single value because the gas is not destroyed over time, but instead moves among different parts of the ocean–atmosphere–land system. Some of the excess carbon dioxide is absorbed quickly (for example, by the ocean surface), but some will remain in the atmosphere for thousands of years, due in part to the very slow process by which carbon is transferred to ocean sediments.

Summary of Key Points

 **U.S. Greenhouse Gas Emissions.** In the United States, greenhouse gas emissions caused by human activities increased by 7 percent from 1990 to 2014. Since 2005, however, total U.S. greenhouse gas emissions have decreased by 7 percent. Electricity generation is the largest source of greenhouse gas emissions in the United States, followed by transportation.

 **Global Greenhouse Gas Emissions.** Worldwide, net emissions of greenhouse gases from human activities increased by 35 percent from 1990 to 2010. Emissions of carbon dioxide, which account for about three-fourths of total emissions, increased by 42 percent over this period.

 **Atmospheric Concentrations of Greenhouse Gases.** Concentrations of carbon dioxide and other greenhouse gases in the atmosphere have increased since the beginning of the industrial era. Almost all of this increase is attributable to human activities.[1] Historical measurements show that the current global atmospheric concentrations of carbon dioxide are unprecedented compared with the past 800,000 years, even after accounting for natural fluctuations.

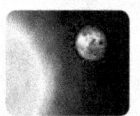 **Climate Forcing.** Climate forcing refers to a change in the Earth's energy balance, leading to either a warming or cooling effect over time. An increase in the atmospheric concentrations of greenhouse gases produces a positive climate forcing, or warming effect. From 1990 to 2015, the total warming effect from greenhouse gases added by humans to the Earth's atmosphere increased by 37 percent. The warming effect associated with carbon dioxide alone increased by 30 percent.

 **U.S. and Global Temperature.** Average temperatures have risen across the contiguous 48 states since 1901. Average global temperatures show a similar trend, and all of the top 10 warmest years on record worldwide have occurred since 1998. Within the United States, temperatures in parts of the North, the West, and Alaska have increased the most.

 **High and Low Temperatures.** Nationwide, unusually hot summer days (highs) have become more common over the last few decades. Unusually hot summer nights (lows) have become more common at an even faster rate. This trend indicates less "cooling off" at night. Although the United States has experienced many winters with unusually low temperatures, unusually cold winter temperatures have become less common—particularly very cold nights (lows).

 **U.S. and Global Precipitation.** Total annual precipitation has increased over land areas in the United States and worldwide. Since 1901, precipitation has increased at an average rate of 0.08 inches per decade over land areas worldwide. However, shifting weather patterns have caused certain areas, such as the Southwest, to experience less precipitation than usual.

 **Heavy Precipitation.** In recent years, a higher percentage of precipitation in the United States has come in the form of intense single-day events. The prevalence of extreme single-day precipitation events remained fairly steady between 1910 and the 1980s but has risen substantially since then. Nationwide, nine of the top 10 years for extreme one-day precipitation events have occurred since 1990.

Tropical Cyclone Activity. Tropical storm activity in the Atlantic Ocean, the Caribbean, and the Gulf of Mexico has increased during the past 20 years. Storm intensity is closely related to variations in sea surface temperature in the tropical Atlantic. However, changes in observation methods over time make it difficult to know for sure whether a longer-term increase in storm activity has occurred.

River Flooding. Increases and decreases in the frequency and magnitude of river flood events vary by region. Floods have generally become larger across parts of the Northeast and Midwest and smaller in the West, southern Appalachia, and northern Michigan. Large floods have become more frequent across the Northeast, Pacific Northwest, and parts of the northern Great Plains, and less frequent in the Southwest and the Rockies.

Drought. Over the period from 2000 through 2015, roughly 20 to 70 percent of the U.S. land area experienced conditions that were at least abnormally dry at any given time. However, this index has not been in use for long enough to compare with historical drought patterns.

A Closer Look: Temperature and Drought in the Southwest. The southwestern United States is particularly sensitive to changes in temperature and thus vulnerable to drought, as even a small decrease in water availability in this already arid region can stress natural systems and further threaten water supplies.

Oceans

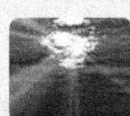

Ocean Heat. Three independent analyses show that the amount of heat stored in the ocean has increased substantially since the 1950s. Ocean heat content not only determines sea surface temperature, but also affects sea level and currents.

Sea Surface Temperature. Ocean surface temperatures increased around the world during the 20th century. Even with some year-to-year variation, the overall increase is clear, and sea surface temperatures have been consistently higher during the past three decades than at any other time since reliable observations began in the late 1800s.

Sea Level. When averaged over all of the world's oceans, sea level has risen at a rate of roughly six-tenths of an inch per decade since 1880. The rate of increase has accelerated in recent years to more than an inch per decade. Changes in sea level relative to the land vary by region. Along the U.S. coastline, sea level has risen the most along the Mid-Atlantic coast and parts of the Gulf coast, where some stations registered increases of more than 8 inches between 1960 and 2015. Sea level has decreased relative to the land in parts of Alaska and the Pacific Northwest.

A Closer Look: Land Loss Along the Atlantic Coast. As sea level rises, dry land and wetlands can turn into open water. Along many parts of the Atlantic coast, this problem is made worse by low elevations and land that is already sinking. Between 1996 and 2011, the coastline from Florida to New York lost more land than it gained.

Coastal Flooding. Flooding is becoming more frequent along the U.S. coastline as sea level rises. Nearly every site measured has experienced an increase in coastal flooding since the 1950s. The rate is accelerating in many locations along the East and Gulf coasts. The Mid-Atlantic region suffers the highest number of coastal flood days and has also experienced the largest increases in flooding.

Ocean Acidity. The ocean has become more acidic over the past few decades because of increased levels of atmospheric carbon dioxide, which dissolves in the water. Higher acidity affects the balance of minerals in the water, which can make it more difficult for certain marine animals to build their protective skeletons or shells.

Arctic Sea Ice. Part of the Arctic Ocean is covered by ice year-round. The area covered by ice is typically smallest in September, after the summer melting season. The annual minimum extent of Arctic sea ice has decreased over time, and in September 2012 it was the smallest ever recorded. The length of the melt season for Arctic ice has grown, and the ice has also become thinner, which makes it more vulnerable to further melting.

Antarctic Sea Ice. Antarctic sea ice extent in September and February has increased somewhat over time. The September maximum extent reached the highest level on record in 2014—about 7 percent larger than the 1981–2010 average. Slight increases in Antarctic sea ice are outweighed by the loss of sea ice in the Arctic during the same time period, however.

Glaciers. Glaciers in the United States and around the world have generally shrunk since the 1960s, and the rate at which glaciers are melting has accelerated over the last decade. The loss of ice from glaciers has contributed to the observed rise in sea level.

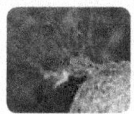

Lake Ice. Lakes in the northern United States are thawing earlier in spring compared with the early 1900s. All 14 lakes studied were found to be thawing earlier in the year, with thaw dates shifting earlier by up to 24 days over the past 110 years.

Community Connection: Ice Breakup in Two Alaskan Rivers. Regions in the far north are warming more quickly than other parts of the world. Two long-running contests on the Tanana and Yukon rivers in Alaska—where people guess the date when the river ice will break up in the spring—provide a century's worth of evidence revealing that the ice on these rivers is generally breaking up earlier in the spring than it once did.

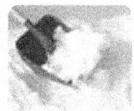

Snowfall. Total snowfall—the amount of snow that falls in a particular location—has decreased in most parts of the country since widespread records began in 1930. One reason for this decline is that nearly 80 percent of the locations studied have seen more winter precipitation fall in the form of rain instead of snow.

Snow Cover. Snow cover refers to the area of land that is covered by snow at any given time. Between 1972 and 2015, the average portion of North America covered by snow decreased at a rate of about 3,300 square miles per year, based on weekly measurements taken throughout the year. There has been much year-to-year variability, however. The length of time when snow covers the ground has become shorter by nearly two weeks since 1972, on average.

Snowpack. The depth of snow on the ground (snowpack) in early spring decreased at more than 90 percent of measurement sites in the western United States between 1955 and 2016. Across all sites, snowpack depth declined by an average of 23 percent during this time period.

Heat-Related Deaths. Since 1979, more than 9,000 Americans were reported to have died as a direct result of heat-related illnesses such as heat stroke. The annual death rate is higher when accounting for deaths in which heat was reported as a contributing factor, including the inter-action of heat and cardiovascular disease. People aged 65+ are a particular concern: a growing demographic group that is several times more likely to die from heat-related cardiovascular disease than the general population. Considerable year-to-year variability and certain limitations of the underlying data for this indicator make it difficult to determine whether the United States has experienced long-term trends in the number of deaths classified as "heat-related."

Heat-Related Illnesses. From 2001 to 2010, a total of about 28,000 heat-related hospitaliza-tions were recorded across 20 states. Annual heat-related hospitalization rates ranged from fewer than one case per 100,000 people in some states to nearly four cases per 100,000 in others. People aged 65+ accounted for more heat-related hospitalizations than any other age group from 2001 to 2010, and males were hospitalized for heat-related illnesses more than twice as often as females.

Heating and Cooling Degree Days. Heating and cooling degree days measure the difference between outdoor temperatures and the temperatures that people find comfortable indoors. As the U.S. climate has warmed in recent years, heating degree days have decreased and cooling degree days have increased overall, suggesting that Americans need to use less energy for heating and more energy for air conditioning.

Lyme Disease. Lyme disease is a bacterial illness spread by ticks that bite humans. Tick habitat and populations are influenced by many factors, including climate. Nationwide, the rate of reported cases of Lyme disease has approximately doubled since 1991. The number and distribu-tion of reported cases of Lyme disease have increased in the Northeast and upper Midwest over time, driven by multiple factors.

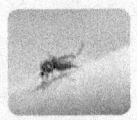

West Nile Virus. West Nile virus is spread by mosquitoes, whose habitat and populations are influenced by temperature and water availability. The incidence of West Nile virus neuroinvasive disease in the United States has varied widely from year to year and among geographic regions since tracking began in 2002. Variation in disease incidence is affected by climate and many other factors, and no obvious long-term trend can be detected yet through this limited data set.

Length of Growing Season. The length of the growing season for crops has increased in almost every state. States in the Southwest (e.g., Arizona and California) have seen the most dramatic increase. In contrast, the growing season has actually become shorter in a few southeastern states. The observed changes reflect earlier spring warming as well as later arrival of fall frosts.

Ragweed Pollen Season. Warmer temperatures and later fall frosts allow ragweed plants to produce pollen later into the year, potentially prolonging the allergy season for millions of people. The length of ragweed pollen season has increased at 10 out of 11 locations studied in the central United States and Canada since 1995. The change becomes more pronounced from south to north.

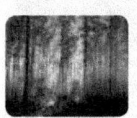 **Wildfires**. Of the 10 years with the largest acreage burned since 1983, nine have occurred since 2000. Fires burn more land in the western United States than in the East.

 **Streamflow.** Changes in temperature, precipitation, snowpack, and glaciers can affect the rate of streamflow and the timing of peak flow. Over the last 75 years, minimum, maximum, and average flows have changed in many parts of the country—some higher, some lower. Most of the rivers and streams measured show peak winter-spring runoff happening at least five days earlier than it did in the mid-20th century.

 **Stream Temperature**. Stream temperatures have risen throughout the Chesapeake Bay region— the area of focus for this indicator. From 1960 through 2014, water temperature increased at 79 percent of the stream sites measured in the region. Temperature has risen by an average of 1.2°F across all sites and 2.2°F at the sites where trends were statistically significant.

 **Tribal Connection: Water Temperature in the Snake River.** Between 1960 and 2015, water temperatures increased by 1.4°F in the Snake River at a site in eastern Washington. Several species of salmon use the Snake River to migrate and spawn, and these salmon play an important role in the diet, culture, religion, and economy of the region's Native Americans.

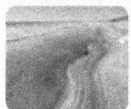 **Great Lakes Water Levels.** Water levels in most of the Great Lakes appear to have declined in the last few decades. However, the most recent levels are all within the range of historical variation. Water levels in lakes are influenced by water temperature, which affects evaporation rates and ice formation.

 **Bird Wintering Ranges.** Some birds shift their range or alter their migration habits to adapt to changes in temperature or other environmental conditions. Long-term studies have found that bird species in North America have shifted their wintering grounds northward by an average of more than 40 miles since 1966, with several species shifting by hundreds of miles.

 **Marine Species Distribution**. The average center of biomass for 105 marine fish and invertebrate species along U.S. coasts shifted northward by about 10 miles between 1982 and 2015. These species also moved an average of 20 feet deeper. Shifts have occurred among several economically important fish and shellfish species. For example, American lobster, black sea bass, and red hake in the Northeast have moved northward by an average of 119 miles.

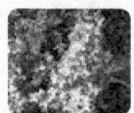 **Leaf and Bloom Dates.** Leaf growth and flower blooms are examples of natural events whose timing can be influenced by climate change. Observations of lilacs and honeysuckles in the contiguous 48 states suggest that first leaf dates and bloom dates show a great deal of year-to-year variability.

 **Community Connection: Cherry Blossom Bloom Dates in Washington, D.C.** Peak bloom dates of the iconic cherry trees in Washington, D.C., recorded since the 1920s, indicate that cherry trees are blooming slightly earlier than in the past. Bloom dates are key to planning the Cherry Blossom Festival, one of the region's most popular spring attractions.

Greenhouse Gases

Greenhouse gases from human activities are the most significant driver of observed climate change since the mid-20th century.[1] The indicators in this chapter characterize emissions of the major greenhouse gases resulting from human activities, the concentrations of these gases in the atmosphere, and how emissions and concentrations have changed over time. When comparing emissions of different gases, these indicators use a concept called "global warming potential" to convert amounts of other gases into carbon dioxide equivalents.

WHY DOES IT MATTER?

As greenhouse gas emissions from human activities increase, they build up in the atmosphere and warm the climate, leading to many other changes around the world—in the atmosphere, on land, and in the oceans. The indicators in other chapters of this report illustrate many of these changes. Such changes have both positive and negative effects on people, society, and the environment—including plants and animals. Because many of the major greenhouse gases stay in the atmosphere for tens to thousands of years after being released, their warming effects on the climate persist over a long time and can therefore affect both present and future generations.

U.S. Greenhouse Gas Emissions

This indicator describes emissions of greenhouse gases in the United States.

A number of factors influence the quantities of greenhouse gases released into the atmosphere, including economic activity, population, consumption patterns, energy prices, land use, and technology. There are several ways to track these emissions, such as by measuring emissions directly, calculating emissions based on the amount of fuel that people burn, and estimating other activities and their associated emissions.

U.S. Greenhouse Gas Emissions and Sinks by Economic Sector, 1990–2014

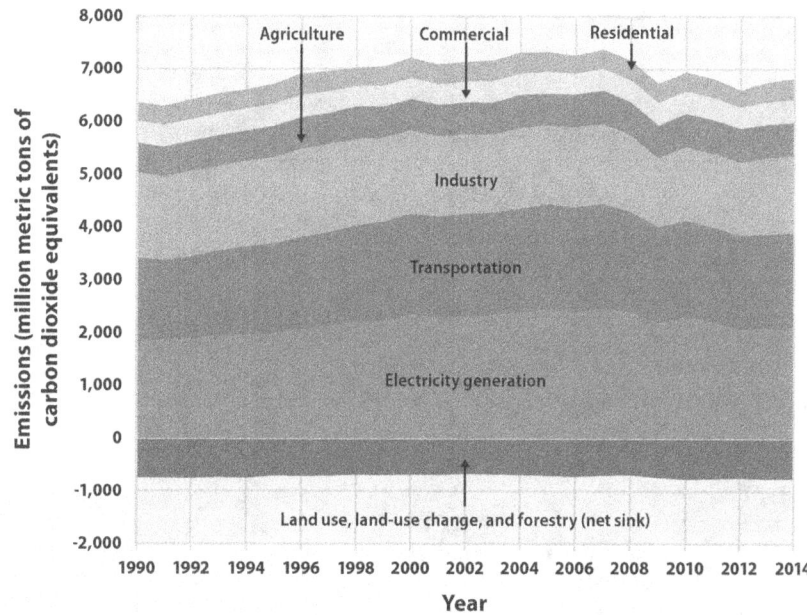

This figure shows greenhouse gas emissions (positive values) and sinks (negative values), by source, in the United States from 1990 to 2014. For consistency, emissions are expressed in million metric tons of carbon dioxide equivalents. All electric power emissions are grouped together in the "Electricity generation" sector, so other sectors such as "Residential" and "Commercial" show only non-electric sources, such as burning oil or gas for heating. The economic sectors shown here do not include emissions from U.S. territories outside the 50 states. Data source: U.S. EPA, 2016[2]

WHAT'S HAPPENING

- In 2014, U.S. greenhouse gas emissions totaled 6,870 million metric tons (15.1 trillion pounds) of carbon dioxide equivalents. This 2014 total represents a 7-percent increase since 1990 but a 7-percent decrease since 2005.

- Among the various sectors of the U.S. economy, electricity generation (power plants) accounts for the largest share of emissions—31 percent of total greenhouse gas emissions since 1990. Transportation is the second-largest sector, accounting for 26 percent of emissions since 1990.

- Emissions sinks, the opposite of emissions sources, absorb carbon dioxide from the atmosphere. In 2014, 11 percent of U.S. greenhouse gas emissions were offset by net sinks resulting from land use and forestry practices. Growing forests remove carbon from the atmosphere, outweighing emissions from wildfires. Other carbon emissions and sinks result from crop practices, burning biofuels, or depositing yard trimmings and food scraps in landfills.

ABOUT THE INDICATOR

This indicator focuses on emissions of carbon dioxide, methane, nitrous oxide, and several fluorinated gases. Data and analysis for this indicator come from EPA's annual *Inventory of U.S. Greenhouse Gas Emissions and Sinks*.[3] This indicator focuses on emissions associated with human activities, though some emissions and sinks from unmanaged lands are also included. Each greenhouse gas has a different lifetime (how long it stays in the atmosphere) and a different ability to trap heat in our atmosphere. To allow different gases to be compared and added together, emissions are converted into carbon dioxide equivalents using each gas's 100-year global warming potential. This analysis uses global warming potentials from the Intergovernmental Panel on Climate Change's Fourth Assessment Report. It starts in 1990, which is a common baseline year for global agreements to track and reduce greenhouse gas emissions. Other parts of this indicator available online track U.S. emissions by greenhouse gas, per capita, and per dollar of gross domestic product since 1990.

Global Greenhouse Gas Emissions

This indicator describes emissions of greenhouse gases worldwide.

ncreasing emissions of greenhouse gases due to human activities worldwide have led to a substantial increase in atmospheric concentrations of these gases (see the Atmospheric Concentrations of Greenhouse Gases indicator on p. 15). Every country around the world emits greenhouse gases into the atmosphere, meaning the root cause of climate change is truly global in scope. Some countries produce far more greenhouse gases than others, and several factors—such as economic activity, population, income level, land use, and climatic conditions—can influence a country's emissions levels. Tracking greenhouse gas emissions worldwide provides a global context for understanding the United States' and other nations' roles in climate change.

ABOUT THE INDICATOR

Data and analysis for this indicator come from the World Resources Institute's Climate Analysis Indicators Tool, which compiles data from peer-reviewed and internationally recognized greenhouse gas inventories developed by EPA and other government agencies worldwide. The Climate Analysis Indicators Tool includes estimates of emissions and sinks associated with land use and forestry activities, which come from global estimates compiled by the Food and Agriculture Organization of the United Nations. Each greenhouse gas has a different lifetime (how long it stays in the atmosphere) and a different ability to trap heat in our atmosphere. To allow different gases to be compared and added together, emissions are converted into carbon dioxide equivalents using each gas's global warming potential, which measures how much a given amount of the gas is estimated to contribute to global warming over a period of 100 years after being emitted. This analysis uses global warming potentials from the Intergovernmental Panel on Climate Change's Second Assessment Report. Other parts of this indicator available online track global greenhouse gas emissions by sector and region since 1990.

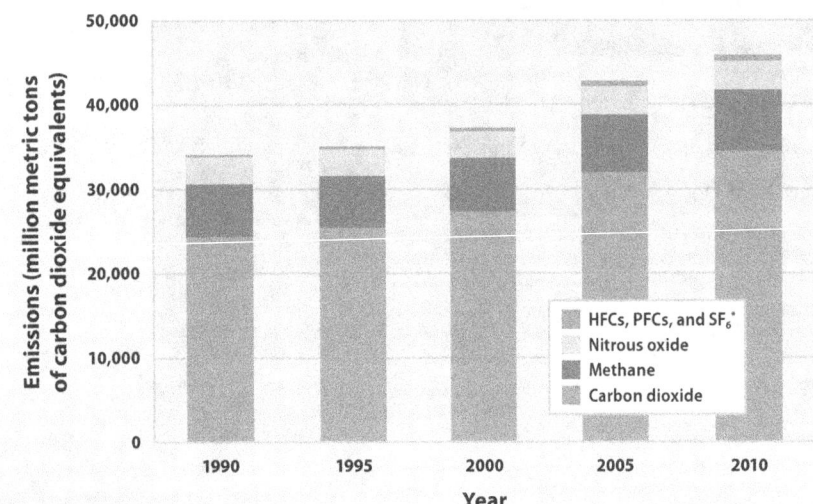

Global Greenhouse Gas Emissions by Gas, 1990–2010

Legend:
- HFCs, PFCs, and SF₆*
- Nitrous oxide
- Methane
- Carbon dioxide

This figure shows worldwide emissions of carbon dioxide, methane, nitrous oxide, and several fluorinated gases from 1990 to 2010. For consistency, emissions are expressed in million metric tons of carbon dioxide equivalents. These totals include emissions and sinks due to land-use change and forestry. Data sources: WRI, 2014;[4] FAO, 2014[5]
* HFCs are hydrofluorocarbons, PFCs are perfluorocarbons, and SF₆ is sulfur hexafluoride.

- In 2010, estimated worldwide emissions from human activities totaled nearly 46 billion metric tons of greenhouse gases, expressed as carbon dioxide equivalents. This represents a 35-percent increase from 1990. These numbers represent net emissions, which include the effects of land use and forestry.

- Between 1990 and 2010, global emissions of all major greenhouse gases increased. Net emissions of carbon dioxide increased by 42 percent, which is particularly important because carbon dioxide accounts for about three-fourths of total global emissions. Nitrous oxide emissions increased the least—9 percent—while emissions of methane increased by 15 percent. Emissions of fluorinated gases more than doubled.

Atmospheric Concentrations of Greenhouse Gases

This indicator describes how the levels of major greenhouse gases in the atmosphere have changed over time.

Since the Industrial Revolution began in the 1700s, people have added a substantial amount of heat-trapping greenhouse gases into the atmosphere by burning fossil fuels, cutting down forests, and conducting other activities (see the U.S. and Global Greenhouse Gas Emissions indicators on pp. 13 and 14). Many of these gases remain in the atmosphere for long time periods ranging from a decade to many millennia, which allows them to become well mixed throughout the global atmosphere. As a result of human activities, these gases are entering the atmosphere more quickly than they are being removed by chemical reactions or by emissions sinks, such as the oceans and vegetation, which absorb greenhouse gases from the atmosphere. Thus, their concentrations are increasing, which contributes to global warming.

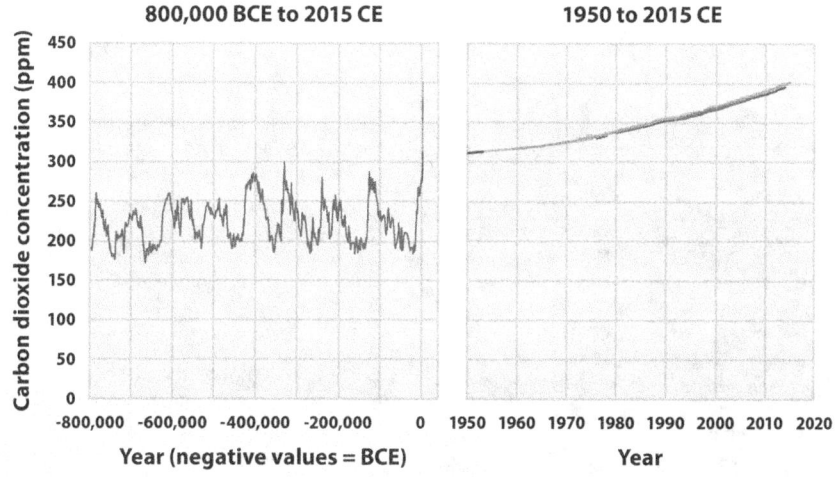

Global Atmospheric Concentrations of Carbon Dioxide Over Time

This figure shows concentrations of carbon dioxide in the atmosphere from hundreds of thousands of years ago through 2015, measured in parts per million (ppm). The data come from a variety of historical ice core studies and recent air monitoring sites around the world. Each line represents a different data source. Data source: Compilation of 10 underlying datasets[6]

WHAT'S HAPPENING

- Global atmospheric concentrations of carbon dioxide have risen significantly over the last few hundred years.

- Historical measurements show that the current global atmospheric concentrations of carbon dioxide are unprecedented compared with the past 800,000 years.

- Since the beginning of the industrial era, concentrations of carbon dioxide have increased from an annual average of 280 ppm in the late 1700s to 401 ppm as measured at Mauna Loa in 2015—a 43-percent increase. This increase is due to human activities.[7]

ABOUT THE INDICATOR

This indicator describes concentrations of greenhouse gases in the atmosphere. The graph above focuses on carbon dioxide, which accounts for the largest share of warming associated with human activities. Recent measurements come from monitoring stations around the world, while measurements of older air come from air bubbles trapped in layers of ice from Antarctica and Greenland. By determining the age of the ice layers and the concentrations of gases trapped inside, scientists can learn what the atmosphere was like thousands of years ago. Other parts of this indicator available online track global atmospheric concentrations of methane and nitrous oxide over the past 800,000 years and global atmospheric concentrations of selected halogenated gases and ozone over the last few decades. Ozone acts as a greenhouse gas in the lower atmosphere.

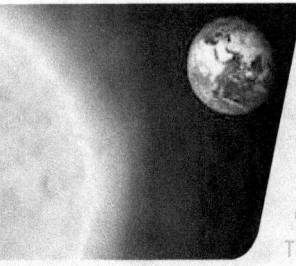

Climate Forcing

This indicator measures the "radiative forcing" or heating effect caused by greenhouse gases in the atmosphere.

When energy from the sun reaches the Earth, the planet absorbs some of this energy and radiates the rest back to space as heat. A variety of physical and chemical factors—some natural and some influenced by humans—can shift the balance between incoming and outgoing energy, which forces changes in the Earth's climate. These changes are measured by the amount of warming or cooling they can produce, which is called "radiative forcing." Changes that have a warming effect are called "positive" forcing, while those that have a cooling effect are called "negative" forcing. When positive and negative forces are out of balance, the result is a change in the Earth's average surface temperature. Greenhouse gases trap heat in the lower atmosphere and cause positive radiative forcing.

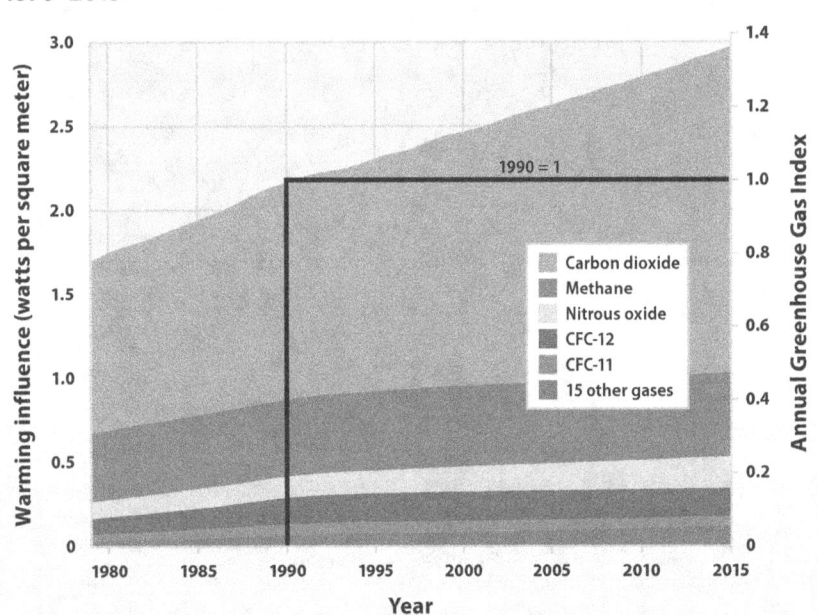

Radiative Forcing Caused by Major Long-Lived Greenhouse Gases, 1979–2015

Legend:
- Carbon dioxide
- Methane
- Nitrous oxide
- CFC-12
- CFC-11
- 15 other gases

This figure shows the amount of radiative forcing caused by various greenhouse gases, based on the change in concentration of these gases in the Earth's atmosphere since 1750. Radiative forcing is calculated in watts per square meter, which represents the size of the energy imbalance in the atmosphere. On the right side of the graph, radiative forcing has been converted to the Annual Greenhouse Gas Index, which is set to a value of 1.0 for 1990. Data source: NOAA, 2016[8]

- In 2015, the Annual Greenhouse Gas Index was 1.37, which represents a 37-percent increase in radiative forcing (a net warming influence) since 1990.

- Of the greenhouse gases shown in the figure, carbon dioxide accounts for by far the largest share of radiative forcing since 1990, and its contribution continues to grow at a steady rate. Carbon dioxide alone would account for a 30-percent increase in radiative forcing since 1990.

- Although the overall Annual Greenhouse Gas Index continues to rise, the rate of increase has slowed somewhat since the baseline year 1990, in large part because methane concentrations have increased at a slower rate in recent years and because chlorofluorocarbon (CFC) concentrations have been declining as production of CFCs has been phased out globally.

ABOUT THE INDICATOR

This indicator measures the average total radiative forcing of 20 long-lived greenhouse gases, including carbon dioxide, methane, and nitrous oxide. The results were calculated by the National Oceanic and Atmospheric Administration based on measured concentrations of these gases in the atmosphere, compared with the concentrations that were present around 1750, before the Industrial Revolution began. Because each gas has a different ability to absorb and emit energy, this indicator converts the changes in greenhouse gas concentrations into a measure of the total radiative forcing (warming effect) caused by each gas. The right side of the graph shows the Annual Greenhouse Gas Index, which compares the radiative forcing for a particular year with the radiative forcing in 1990, which is a common baseline year for global agreements to track and reduce greenhouse gas emissions.

 # Weather and Climate

Rising global average temperature is associated with widespread changes in weather patterns. Scientific studies indicate that extreme weather events such as heat waves and large storms are likely to become more frequent or more intense with human-induced climate change. This chapter focuses on observed changes in temperature, precipitation, storms, floods, and droughts.

WHY DOES IT MATTER?

Long-term changes in climate can directly or indirectly affect many aspects of society in potentially disruptive ways. For example, warmer average temperatures could increase air conditioning costs and affect the spread of diseases like Lyme disease, but could also improve conditions for growing some crops. More extreme variations in weather are also a threat to society. More frequent and intense extreme heat events can increase illnesses and deaths, especially among vulnerable populations, and damage some crops. While increased precipitation can replenish water supplies and support agriculture, intense storms can damage property; cause loss of life and population displacement; and temporarily disrupt essential services such as transportation, telecommunications, energy, and water supplies.

WEATHER AND CLIMATE

Weather is the state of the atmosphere at any given time and place. Most of the weather that affects people, agriculture, and ecosystems takes place in the lower layer of the atmosphere. Familiar aspects of weather include temperature, precipitation, clouds, and wind that people experience throughout the course of a day. Severe weather conditions include hurricanes, tornadoes, blizzards, and droughts.

Climate is the long-term average of the weather in a given place. While the weather can change in minutes or hours, a change in climate is something that develops over longer periods of decades to centuries. Climate is defined not only by average temperature and precipitation but also by the type, frequency, duration, and intensity of weather events such as heat waves, cold spells, storms, floods, and droughts.

While the concepts of climate and weather are often confused, it is important to understand the difference. For example, the eastern United States experienced a cold and snowy winter in 2014/2015, but this short-term regional weather phenomenon does not negate the long-term rise in national and global temperatures, sea level, or other climate indicators.

U.S. and Global Temperature

This indicator describes trends in average surface temperature for the United States and the world.

Warmer temperatures are one of the most direct signs that the climate is changing. Concentrations of heat-trapping greenhouse gases are increasing in the Earth's atmosphere (see the Atmospheric Concentrations of Greenhouse Gases indicator on p. 15). In response, average temperatures at the Earth's surface are increasing and are expected to continue rising. Because climate change can shift the wind patterns and ocean currents that drive the world's climate system, however, some areas are warming more than others, and some have experienced cooling.

ABOUT THE INDICATOR

This indicator is based on daily temperature records from thousands of long-term weather monitoring stations, which have been compiled by the National Oceanic and Atmospheric Administration's National Centers for Environmental Information. The indicator was developed by calculating annual anomalies, or differences, compared with the average temperature from 1901 to 2000. For example, an anomaly of +2.0 degrees means the average temperature was 2 degrees higher than the long-term average. Daily, monthly, and annual anomalies have been calculated for each weather station. Global anomalies have been determined by dividing the world into a grid, averaging the data for each cell of the grid, and then averaging the grid cells together. For the map, anomalies have been averaged together and compared over time within small regions called climate divisions. The online version of this indicator also includes a graph of annual temperature anomalies for the contiguous 48 states since 1901. Hawaii and U.S. territories are not included, due to limitations in available data.

WHAT'S HAPPENING

- Worldwide, 2015 was the warmest year on record and 2006–2015 was the warmest decade on record since thermometer-based observations began. Global average surface temperature has risen at an average rate of 0.15°F per decade since 1901.

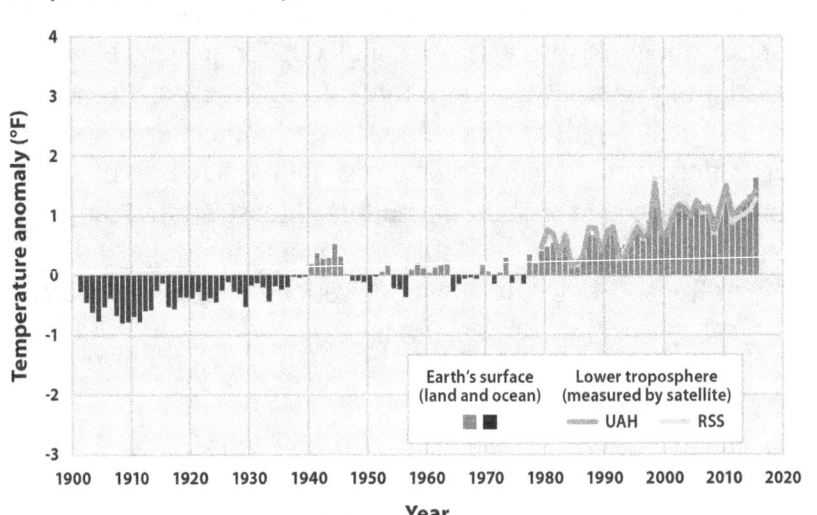

Temperatures Worldwide, 1901–2015

This figure shows how annual average temperatures worldwide have changed since 1901. Surface data come from a combined set of land-based weather stations and sea surface temperature measurements. Satellite measurements cover the lower troposphere, which is the lowest level of the Earth's atmosphere. "UAH" and "RSS" represent two different methods of analyzing the original satellite measurements. This graph uses the 1901–2000 average as a baseline for depicting change. Choosing a different baseline period would not change the shape of the data over time. Data source: NOAA, 2016[1]

Rate of Temperature Change in the United States, 1901–2015

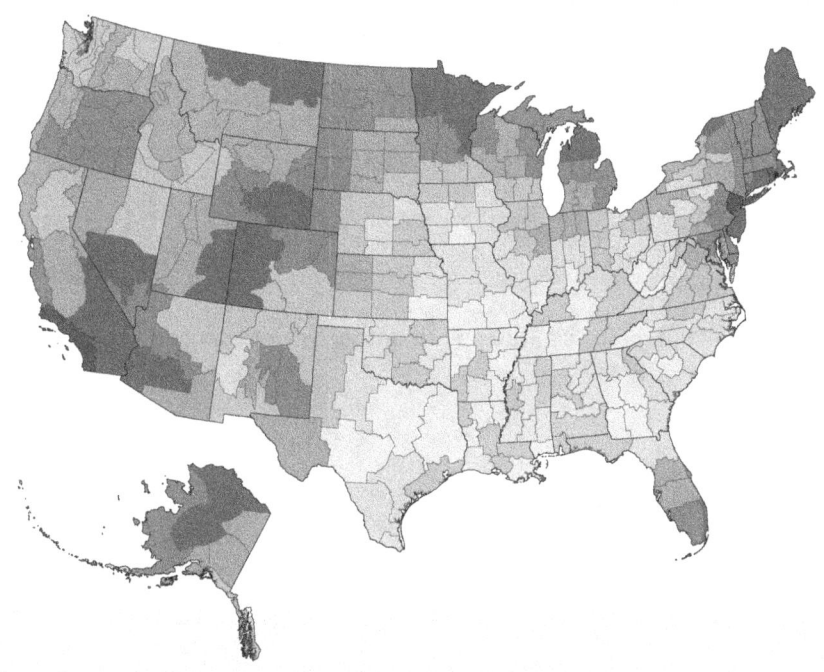

Rate of temperature change (°F per century):

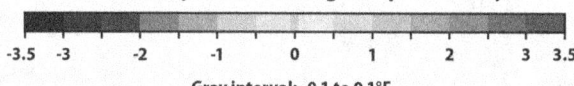

-3.5 -3   -2   -1   0   1   2   3 3.5

Gray interval: -0.1 to 0.1°F

This figure shows how annual average air temperatures have changed in different parts of the United States since the early 20ᵗʰ century (since 1901 for the contiguous 48 states and 1925 for Alaska). The data are shown for climate divisions, as defined by the National Oceanic and Atmospheric Administration. Data source: NOAA, 2016[2]

- The average surface temperature across the contiguous 48 states has risen at an average rate of 0.14°F per decade since 1901, which is similar to the global rate. Since 1979, the contiguous 48 states have warmed by 0.29°F to 0.46°F per decade, which is faster than the global rate.

- Some parts of the United States have experienced more warming than others. The North, the West, and Alaska have seen temperatures increase the most, while some parts of the Southeast have experienced little change. Not all of these regional trends are statistically significant, however.

Weather & Climate

19

High and Low Temperatures

This indicator describes trends in unusually hot and cold temperatures across the United States.

Unusually hot or cold temperatures can result in prolonged extreme weather events like summer heat waves or winter cold spells. Heat waves can lead to illness and death, particularly among older adults, young children, and other populations of concern (see the Heat-Related Deaths and Heat-Related Illnesses indicators on pp. 60 and 62). People can also die from exposure to extreme cold (hypothermia). In addition, exposure to extreme heat and cold can damage crops and injure or kill livestock. Extreme heat can lead to power outages as heavy demands for air conditioning strain the power grid, while extremely cold weather increases the need for heating fuel. Record-setting daily temperatures, heat waves, and cold spells are a natural part of day-to-day variation in weather. As the Earth's climate warms overall, however, heat waves are expected to become more frequent, longer, and more intense, while cold spells are expected to decrease.[3,4]

WHAT'S HAPPENING

- Nationwide, unusually hot summer days (highs) have become more common over the last few decades. Unusually hot summer nights (lows) have become more common at an even faster rate. This trend indicates less "cooling off" at night.

Area of the Contiguous 48 States With Unusually Hot Summer Temperatures, 1910–2015

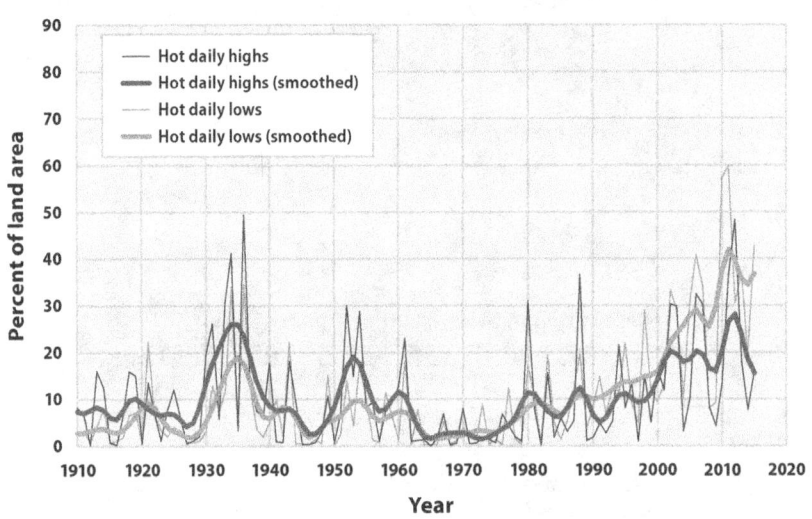

This graph shows the percentage of the land area of the contiguous 48 states with unusually hot daily high and low temperatures during the months of June, July, and August. The thin lines represent individual years, while the thick lines show a nine-year weighted average. Red lines represent daily highs, while orange lines represent daily lows. The term "unusual" in this case is based on the long-term average conditions at each location.
Data source: NOAA, 2015[5]

HEALTH CONNECTION

As extremely hot temperatures become more common, people may be exposed to extreme heat more often. This could increase the risk of heat-related illnesses and deaths—particularly as nighttime temperatures rise and people are less able to cool off at night. Lack of air conditioning, working outdoors, and other social factors can increase exposure among certain groups.[6]

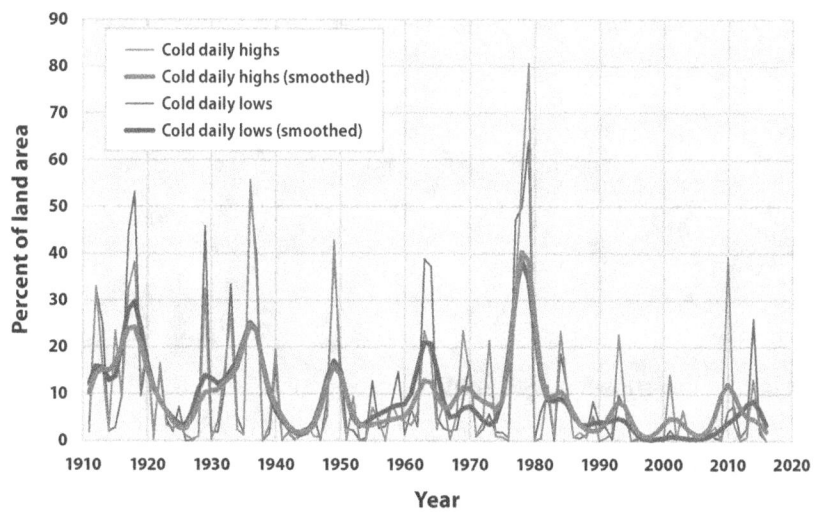

Area of the Contiguous 48 States With Unusually Cold Winter Temperatures, 1911–2016

Legend:
- Cold daily highs
- Cold daily highs (smoothed)
- Cold daily lows
- Cold daily lows (smoothed)

Y-axis: Percent of land area (0 to 90)
X-axis: Year (1910 to 2020)

This graph shows the percentage of the land area of the contiguous 48 states with unusually cold daily high and low temperatures during the months of December, January, and February. The thin lines represent individual years, while the thick lines show a nine-year weighted average. Blue lines represent daily highs, while purple lines represent daily lows. The term "unusual" in this case is based on the long-term average conditions at each location. Data source: NOAA, 2016[7]

- The 20th century had many winters with widespread patterns of unusually low temperatures, including a particularly large spike in the late 1970s. Since the 1980s, though, unusually cold winter temperatures have become less common—particularly very cold nights (lows).

Weather & Climate

ABOUT THE INDICATOR

This indicator is based on temperature measurements from weather stations overseen by the National Oceanic and Atmospheric Administration's National Weather Service. National patterns can be determined by dividing the country into a grid and examining the data for one station in each cell of the grid. This method ensures that the results are not biased toward regions that happen to have many stations close together. The figures show trends in the percentage of the country's area experiencing unusually hot temperatures in the summer and unusually cold temperatures in the winter. These graphs are based on daily maximum temperatures, which usually occur during the day, and daily minimum temperatures, which usually occur at night. At each station, the recorded highs and lows are compared with the full set of historical records. After averaging over a particular month or season of interest, the coldest 10 percent of years are considered "unusually cold" and the warmest 10 percent are "unusually hot." Additional components of this indicator are available online, including a graph of changes in annual heat wave index values, maps showing changes in unusually hot and cold days, and a graph that tracks record daily high and low temperatures for the contiguous 48 states.

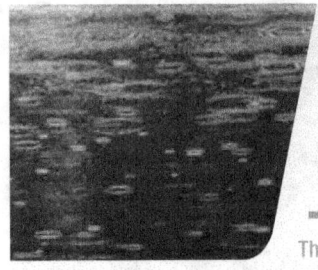

U.S. and Global Precipitation

This indicator describes trends in average precipitation for the United States and the world.

Precipitation can have wide-ranging effects on human well-being and ecosystems. Rainfall, snowfall, and the timing of snowmelt can all affect the amount of surface water and groundwater available for drinking, irrigation, and industry. They also influence river flooding and can determine what types of animals and plants (including crops) can survive in a particular place. Changes in precipitation can disrupt a wide range of natural processes, particularly if these changes occur more quickly than plant and animal species can adapt. As average temperatures at the Earth's surface rise (see the U.S. and Global Temperature indicator on p. 18), more evaporation occurs, which in turn increases overall precipitation. Therefore, a warming climate is expected to increase precipitation in many areas. Just as precipitation patterns vary across the world, however, so will the precipitation effects of climate change. Some areas will experience decreased precipitation. Also, because higher temperatures lead to more evaporation, increased precipitation will not necessarily increase the amount of water available for drinking, irrigation, and industry (see the Drought indicator on p. 28).

WHAT'S HAPPENING

- On average, total annual precipitation has increased over land areas worldwide. Since 1901, global precipitation has increased at an average rate of 0.08 inches per decade.

Precipitation Worldwide, 1901–2015

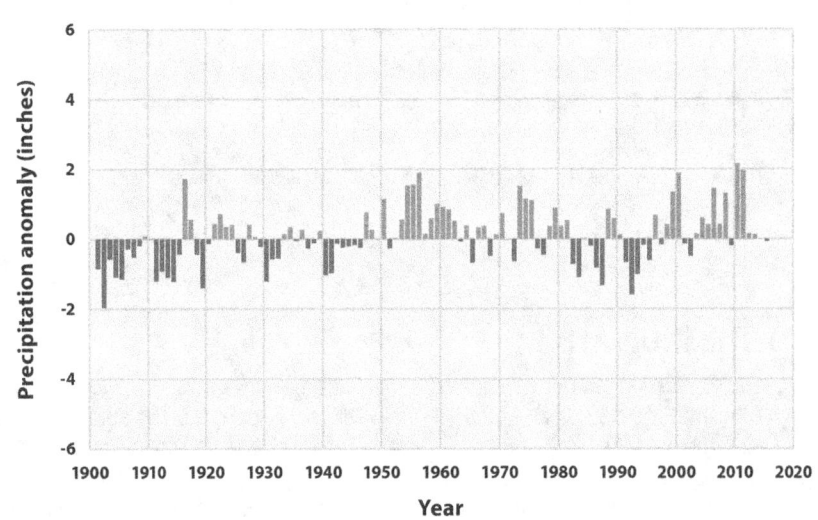

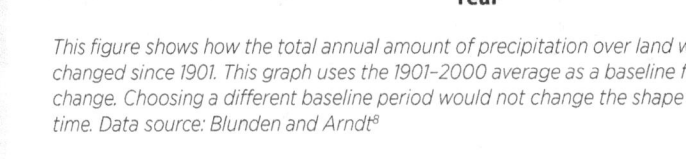

This figure shows how the total annual amount of precipitation over land worldwide has changed since 1901. This graph uses the 1901–2000 average as a baseline for depicting change. Choosing a different baseline period would not change the shape of the data over time. Data source: Blunden and Arndt[8]

Change in Precipitation in the United States, 1901–2015

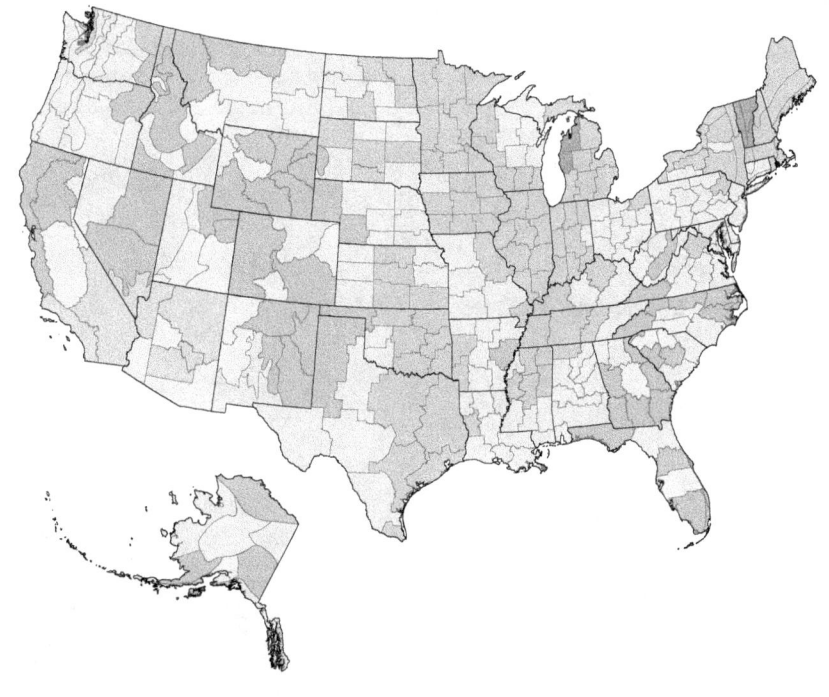

- On average, total annual precipitation has increased over land areas in the United States. Some parts of the United States have experienced greater increases in precipitation than others. A few areas, such as the Southwest, have seen a decrease in precipitation. Not all of these regional trends are statistically significant, however.

Percent change in precipitation:

-30   -20   -10   -2 2   10   20   30

This figure shows the rate of change in total annual precipitation in different parts of the United States since the early 20th century (since 1901 for the contiguous 48 states and 1925 for Alaska). The data are shown for climate divisions, as defined by the National Oceanic and Atmospheric Administration. Data source: NOAA, 2016[9]

ABOUT THE INDICATOR

This indicator is based on daily precipitation records from thousands of long-term weather monitoring stations, which have been compiled by the National Oceanic and Atmospheric Administration's National Centers for Environmental Information. The indicator was developed by looking at total annual precipitation at each weather station, comparing annual totals with long-term (1901–2000) averages to determine annual anomalies (differences), and examining trends in anomalies over time. Global anomalies have been determined by dividing the world into a grid, averaging the data for each cell of the grid, and then averaging the grid cells together. For the map, anomalies have been averaged together and compared over time within small regions called climate divisions. The online version of this indicator also includes a graph of annual precipitation anomalies for the contiguous 48 states since 1901. Hawaii and U.S. territories are not included, due to limitations in available data.

Heavy Precipitation

This indicator tracks the frequency of heavy precipitation events in the United States.

Heavy precipitation refers to instances during which the amount of rain or snow experienced in a location substantially exceeds what is normal. What constitutes a period of heavy precipitation varies according to location and season. Climate change can affect the intensity and frequency of precipitation. Warmer oceans increase the amount of water that evaporates into the air. When more moisture-laden air moves over land or converges into a storm system, it can produce more intense precipitation—for example, heavier rain and snow storms.[10] The potential impacts of heavy precipitation include crop damage, soil erosion, and an increase in flood risk (see the River Flooding indicator on p. 26). In addition, runoff from precipitation can impair water quality as pollutants deposited on land wash into water bodies.

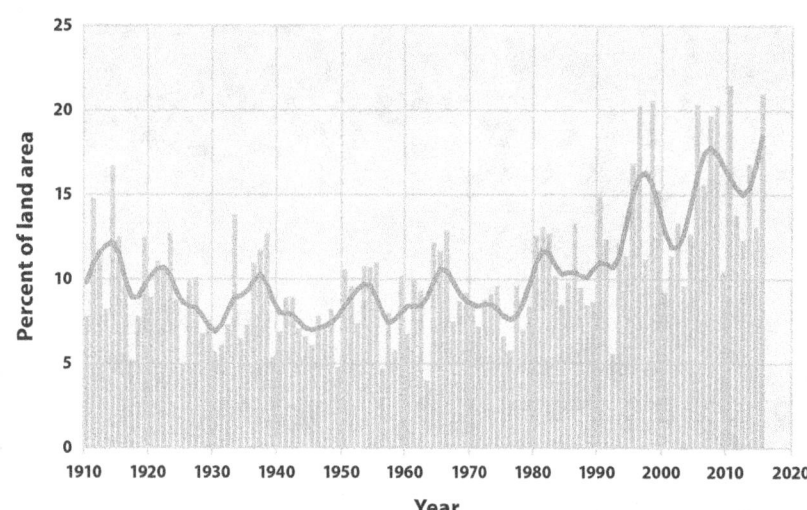

Extreme One-Day Precipitation Events in the Contiguous 48 States, 1910–2015

This figure shows the percentage of the land area of the contiguous 48 states where a much greater than normal portion of total annual precipitation has come from extreme single-day precipitation events. The bars represent individual years, while the line is a nine-year weighted average.
Data source: NOAA, 2016[11]

WHAT'S HAPPENING

• In recent years, a larger percentage of precipitation has come in the form of intense single-day events. Nine of the top 10 years for extreme one-day precipitation events have occurred since 1990.

• The prevalence of extreme single-day precipitation events remained fairly steady between 1910 and the 1980s, but has risen substantially since then. Over the entire period from 1910 to 2015, the portion of the country experiencing extreme single-day precipitation events increased at a rate of about half a percentage point per decade.

HEALTH CONNECTION

Heavy precipitation events followed by extreme flooding events can lead to injuries and even drownings. Flooding can also damage buildings, allowing water or moisture to enter. This could lead to mold, bacteria, or other air quality problems that have adverse effects on health, such as worsening of asthma.[12]

ABOUT THE INDICATOR

This indicator is based on precipitation measurements collected at weather stations throughout the contiguous 48 states. These data are compiled and managed by the National Oceanic and Atmospheric Administration's National Centers for Environmental Information. Heavy precipitation events can be measured by tracking their frequency, examining their return period (the chance that the event will be equaled or exceeded in a given year), or directly measuring the amount of precipitation in a certain period (for example, inches of rain falling in a 24-hour period). One way to track heavy precipitation is by calculating what percentage of a particular location's total precipitation in a given year has come in the form of extreme one-day events—or, in other words, what percentage of precipitation is arriving in short, intense bursts, as shown here. The results shown here are consistent with other methods of assessing changes in heavy precipitation, which also show increases over time.[13] The online version of this indicator also tracks unusually high annual precipitation totals in the contiguous 48 states since 1895 using a scale called the Standardized Precipitation Index.

Tropical Cyclone Activity

This indicator examines the frequency, intensity, and duration of hurricanes and other tropical storms in the Atlantic Ocean, Caribbean, and Gulf of Mexico.

Hurricanes, tropical storms, and other intense rotating storms fall into a general category called cyclones. The effects of tropical cyclones are numerous and well known. At sea, storms disrupt and endanger shipping traffic. When cyclones encounter land, their intense rains and high winds can cause severe property damage, loss of life, soil erosion, and flooding. The associated storm surge—the large volume of ocean water pushed toward shore by the cyclone's strong winds—can cause severe flooding and destruction. Climate change is expected to affect tropical cyclones by increasing sea surface temperatures, a key factor that influences cyclone formation and behavior. The U.S. Global Change Research Program and the Intergovernmental Panel on Climate Change project that, more likely than not, tropical cyclones will become more intense over the 21st century, with higher wind speeds and heavier rains.[14, 15]

WHAT'S HAPPENING

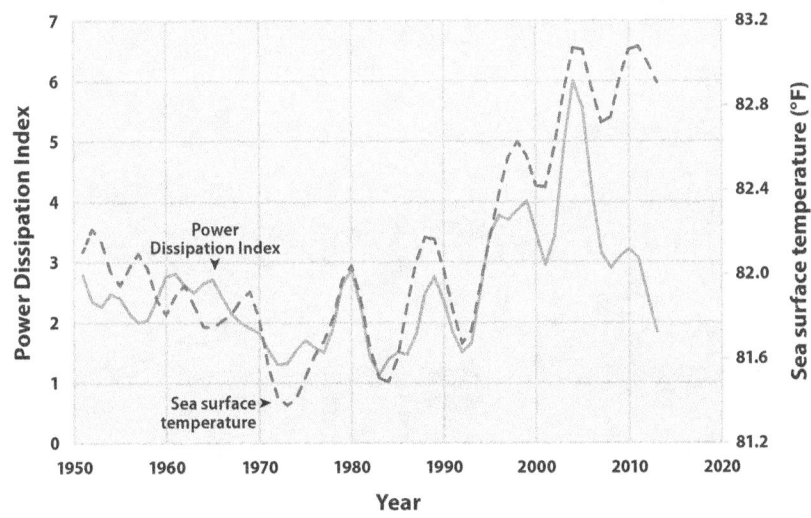

North Atlantic Tropical Cyclone Activity According to the Power Dissipation Index, 1949–2015

This figure presents annual values of the Power Dissipation Index, which accounts for cyclone strength, duration, and frequency. Tropical North Atlantic sea surface temperature trends are provided for reference. Note that sea surface temperature is measured in different units, but the values have been plotted alongside the Power Dissipation Index to show how they compare. The lines have been smoothed using a five-year weighted average, plotted at the middle year. The most recent average (2011–2015) is plotted at 2013.
Data source: Emanuel, 2016[16]

- The Power Dissipation Index shows fluctuating cyclone intensity for most of the mid- to late 20th century, followed by a noticeable increase since 1995. These trends are shown with associated variations in sea surface temperature in the tropical North Atlantic, for comparison.

- Despite the apparent increases in tropical cyclone activity during recent years, changes in observation methods over time make it difficult to know whether tropical storm activity has actually shown a longer-term increase.[17]

ABOUT THE INDICATOR

This indicator is based on data maintained by the National Oceanic and Atmospheric Administration's National Hurricane Center in a database referred to as HURDAT (HURricane DATa). It presents an analysis of HURDAT data using the Power Dissipation Index developed by Dr. Kerry Emanuel at the Massachusetts Institute of Technology. This index tracks the frequency, strength, and duration of tropical cyclones, based on measurements of wind speed. Other parts of this indicator available online track long-term trends in the number of hurricanes in the North Atlantic Ocean, as well as tropical cyclone activity according to another index called the Accumulated Cyclone Energy Index.

River Flooding

Rivers and streams experience flooding as a natural result of large rain storms or spring snowmelt that quickly drains into streams and rivers. Climate change may cause these floods to become larger or more frequent than they used to be in some places, yet smaller and less frequent in other places. Warmer temperatures can lead to changes in the size and frequency of heavy precipitation events, which may in turn affect the size and frequency of river flooding (see the Heavy Precipitation indicator on p. 24).[18] Changes in streamflow, the timing of snowmelt (see the Streamflow indicator on p. 74), and the amount of snowpack that accumulates in the winter (see the Snowpack indicator on p. 52) can also affect flood patterns. Although regular flooding helps to maintain the nutrient balance of soils in the flood plain, larger or more frequent floods could damage homes, roads, bridges, and other infrastructure; wipe out farmers' crops; harm or displace people; contaminate water supplies; and disrupt ecosystems by displacing aquatic life, impairing water quality, and increasing soil erosion.

WHAT'S HAPPENING

- Floods have generally become larger in rivers and streams across large parts of the Northeast and Midwest. Flood magnitude has generally decreased in the West, southern Appalachia, and northern Michigan.

Change in the Magnitude of River Flooding in the United States, 1965–2015

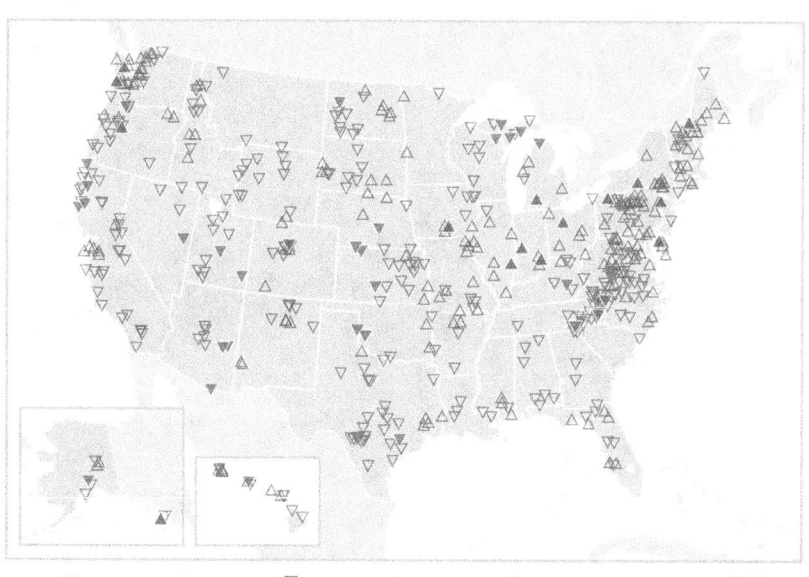

▼ **Significant decrease**   ▽ **Insignificant decrease**   △ **Insignificant increase**   ▲ **Significant increase**

This figure shows changes in the size of flooding events in rivers and streams in the United States between 1965 and 2015. Blue upward-pointing symbols show locations where floods have become larger; brown downward-pointing symbols show locations where floods have become smaller. Larger, solid-color symbols represent stations where the change was statistically significant. Data source: Slater and Villarini, 2016[19]

Change in the Frequency of River Flooding in the United States, 1965–2015

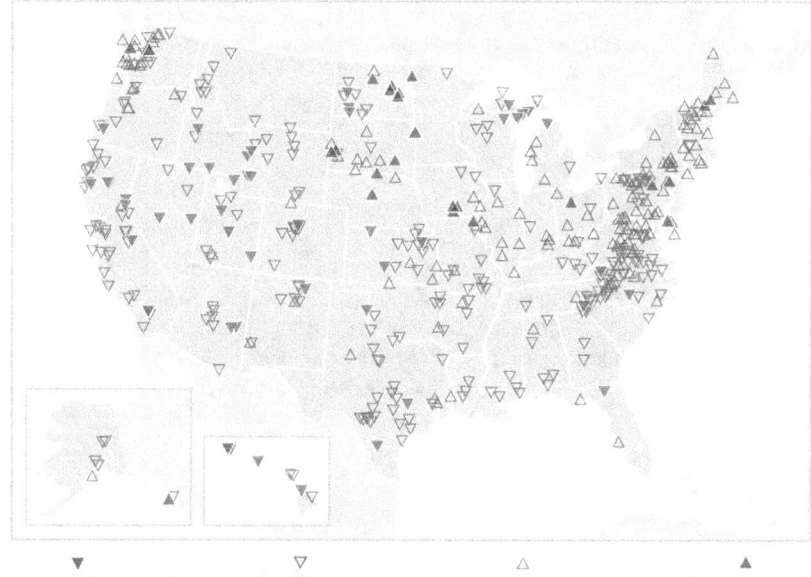

▼	▽	△	▲
Significant decrease	**Insignificant decrease**	**Insignificant increase**	**Significant increase**

This figure shows changes in the frequency of flooding events in rivers and streams in the United States between 1965 and 2015. Blue upward-pointing symbols show locations where floods have become more frequent; brown downward-pointing symbols show locations where floods have become less frequent. Larger, solid-color symbols represent stations where the change was statistically significant. Data source: Slater and Villarini, 2016[20]

- Large floods have become more frequent across the Northeast, Pacific Northwest, and parts of the northern Great Plains. Flood frequency has decreased in some other parts of the country, especially the Southwest and the Rockies.

- Increases and decreases in frequency and magnitude of river flood events generally coincide with increases and decreases in the frequency of heavy rainfall events.[21,22]

Weather & Climate

HEALTH CONNECTION

In addition to the physical health impacts of weather-related disasters, like drowning, injuries, or infections, some people exposed to floods also experience serious mental health consequences. Extreme flood events that involve loss of life or homes are associated with long-term anxiety, depression, and post-traumatic stress disorder.[23]

ABOUT THE INDICATOR

This indicator is based on data from stream gauges maintained by the U.S. Geological Survey. Each gauge measures water level and discharge—the amount of water flowing past the gauge. This indicator uses peak and daily discharge data from a subset of long-term stream gauge stations to identify when the largest flow events have happened and how the size and frequency of large flood events have changed over time. Besides climate change, several other types of human influences could affect the frequency and magnitude of floods—for example, dams, floodwater management activities, agricultural practices, and changes in land use. To minimize these influences, this indicator focuses on a set of sites that are not heavily influenced by human activities. The analysis was developed and updated by researchers at the University of Iowa.

Drought

This indicator measures drought conditions of U.S. lands.

Meteorologists generally define drought as a prolonged period of dry weather caused by a lack of precipitation that results in a serious water shortage for some activity, population, or ecological system. Drought can also be thought of as an extended imbalance between precipitation and evaporation. As average temperatures have risen due to climate change, evaporation has increased, making more water available in the air for precipitation, but contributing to drying over some land areas and less moisture in the soil. Drought conditions can negatively affect agriculture, water supplies, energy production, and many other aspects of society. Lower streamflow and groundwater levels can also harm plants and animals, and dried-out vegetation increases the risk of wildfires.

ABOUT THE INDICATOR

Drought can be measured by looking at precipitation, soil moisture, streamflow, vegetation health, and other variables.[24] The most widely used method is the Palmer Drought Severity Index, which is calculated from precipitation and temperature measurements at weather stations. The Palmer Index is shown in the graph below, based on data from the National Oceanic and Atmospheric Administration. The second graph shows a newer index called the Drought Monitor, which is based on a combination of drought indices (including Palmer) plus additional factors such as snow water content, groundwater levels, reservoir storage, pasture/range conditions, and other impacts. The Drought Monitor uses codes from D0 to D4 to classify drought severity. Drought Monitor data were provided by the National Drought Mitigation Center.

WHAT'S HAPPENING

- Average drought conditions across the nation have varied since records began in 1895. The 1930s and 1950s saw the most widespread droughts, while the last 50 years have generally been wetter than average.

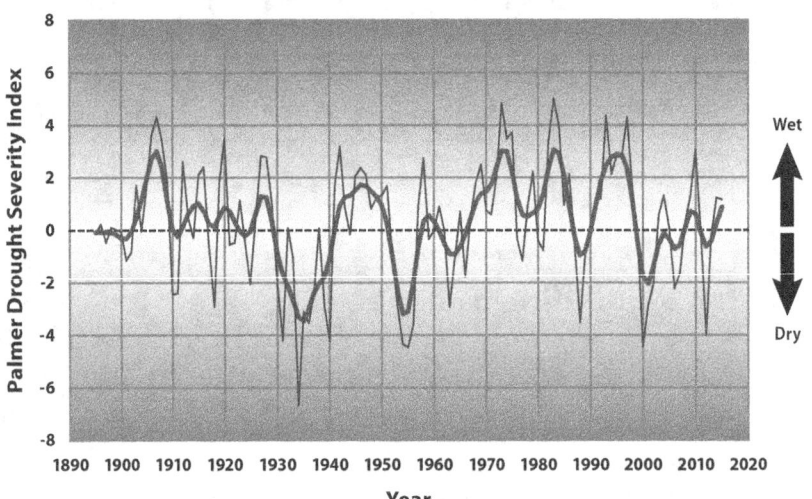

Average Drought Conditions in the Contiguous 48 States, 1895–2015

This chart shows annual values of the Palmer Drought Severity Index, averaged over the entire area of the contiguous 48 states. Positive values represent wetter-than-average conditions, while negative values represent drier-than-average conditions. A value between -2 and -3 indicates moderate drought, -3 to -4 is severe drought, and -4 or below indicates extreme drought. The thicker line is a nine-year weighted average.
Data source: NOAA, 2016[25]

HEALTH CONNECTION

Rising temperatures and prolonged drought pose unique threats to indigenous populations because of their economic and cultural dependence on land and water supplies. Warming and drought can threaten medicinal and culturally important plants and animals, and can reduce water quality and availability, making tribal populations particularly vulnerable to waterborne illnesses.[26]

U.S. Lands Under Drought Conditions, 2000–2015

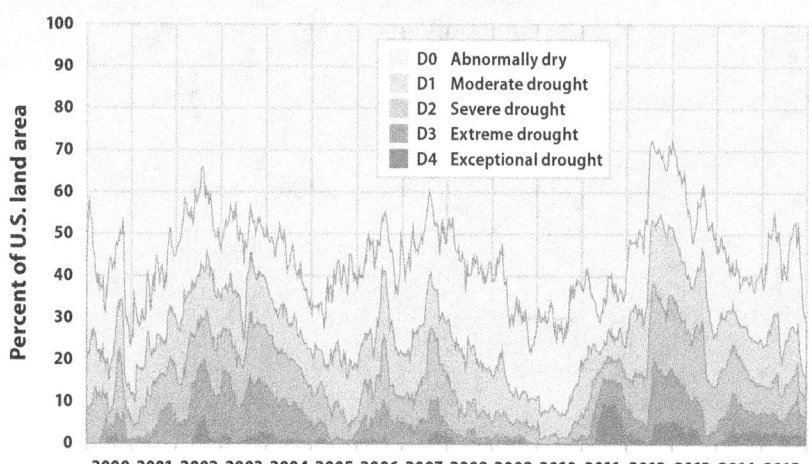

D0 Abnormally dry
D1 Moderate drought
D2 Severe drought
D3 Extreme drought
D4 Exceptional drought

This chart shows the percentage of U.S. lands classified under drought conditions from 2000 through 2015. This figure uses the U.S. Drought Monitor classification system, which is described in the table below. The data cover all 50 states plus Puerto Rico.
Data source: National Drought Mitigation Center, 2016[27]

- Over the period from 2000 through 2015, roughly 20 to 70 percent of the U.S. land area experienced conditions that were at least abnormally dry at any given time. The years 2002–2003 and 2012–2013 had a relatively large area with at least abnormally dry conditions, while 2001, 2005, and 2009–2011 had substantially less area experiencing drought.

- During the latter half of 2012, more than half of the U.S. land area was covered by moderate or greater drought. In several states, 2012 was among the driest years on record.[28] See Temperature and Drought in the Southwest on p. 30 for a closer look at recent drought conditions in one of the hardest-hit regions.

Categories of Drought Severity

Category	Description	Possible Impacts
D0	Abnormally dry	Going into drought: short-term dryness slowing planting or growth of crops or pastures. Coming out of drought: some lingering water deficits; pastures or crops not fully recovered.
D1	Moderate drought	Some damage to crops or pastures; streams, reservoirs, or wells low; some water shortages developing or imminent; voluntary water use restrictions requested.
D2	Severe drought	Crop or pasture losses likely; water shortages common; water restrictions imposed.
D3	Extreme drought	Major crop/pasture losses; widespread water shortages or restrictions.
D4	Exceptional drought	Exceptional and widespread crop/pasture losses; shortages of water in reservoirs, streams, and wells, creating water emergencies.

Experts update the U.S. Drought Monitor weekly and produce maps that illustrate current conditions as well as short- and long-term trends. Major participants include the National Oceanic and Atmospheric Administration, the U.S. Department of Agriculture, and the National Drought Mitigation Center. For a map of current drought conditions, visit the Drought Monitor website at: http://droughtmonitor.unl.edu.

A CLOSER LOOK: DROUGHT IN THE SOUTHWEST

Much of the American Southwest (Arizona, California, Colorado, Nevada, New Mexico, and Utah) experiences low annual rainfall and seasonally high temperatures that contribute to its characteristic arid climate. Yet this landscape actually supports a vast array of plants and animals, along with millions of people who call it home. Water is already scarce, so even a small increase in temperature (which drives evaporation) or a decrease in precipitation can threaten natural systems and society. Droughts also contribute to increased pest outbreaks and wildfires, and they reduce the amount of water available for generating electricity. The last decade has seen the most persistent droughts in the Southwest since recordkeeping began in 1895.

Drought Severity in the Southwestern United States, 1895–2015

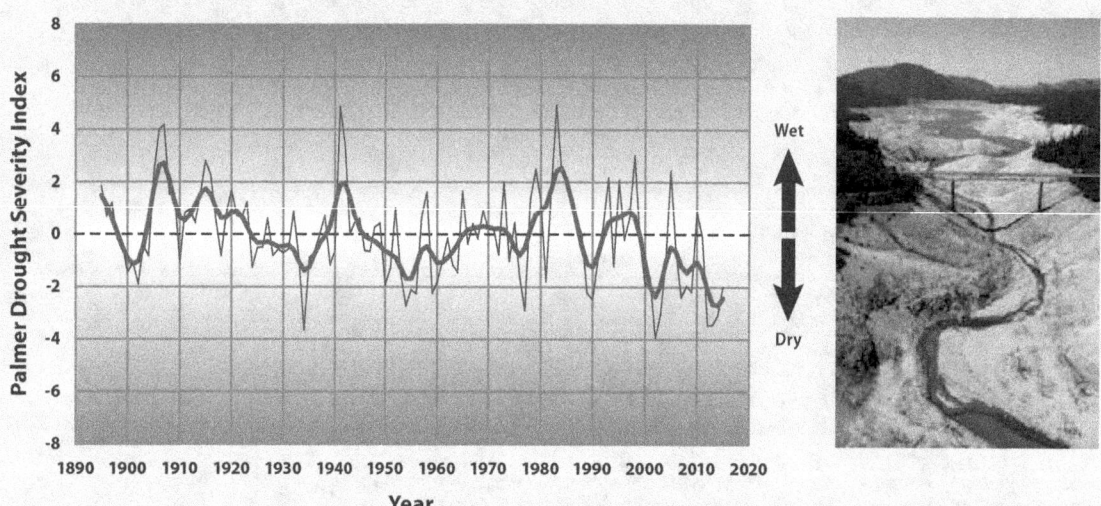

*This chart shows annual values of the Palmer Drought Severity Index, averaged over six states in the Southwest (Arizona, California, Colorado, Nevada, New Mexico, and Utah). Positive values represent wetter-than-average conditions, while negative values represent drier-than-average conditions. A value between -2 and -3 indicates moderate drought, -3 to -4 is severe drought, and -4 or below indicates extreme drought. The thicker line is a nine-year weighted average.
Data source: NOAA, 2016[29]*

 # Oceans

Covering about 70 percent of the Earth's surface, the world's oceans have a two-way relationship with weather and climate. The oceans influence the weather on local to global scales, while changes in climate can fundamentally alter many properties of the oceans. This chapter examines how some of these important characteristics of the oceans have changed over time.

WHY DOES IT MATTER?

As greenhouse gases trap more energy from the sun, the oceans are absorbing more heat, resulting in an increase in sea surface temperatures and rising sea level. Changes in ocean temperatures and currents brought about by climate change will lead to alterations in climate patterns around the world. For example, warmer waters may promote the development of stronger storms in the tropics, which can cause property damage and loss of life. The impacts associated with sea level rise and stronger storms are especially relevant to coastal communities.

Although the oceans help reduce climate change by storing large amounts of carbon dioxide, increasing levels of dissolved carbon are changing the chemistry of seawater and making it more acidic. Increased ocean acidity makes it more difficult for certain organisms, such as corals and shellfish, to build their skeletons and shells. These effects, in turn, could substantially alter the biodiversity and productivity of ocean ecosystems.

Changes in ocean systems generally occur over much longer time periods than in the atmosphere, where storms can form and dissipate in a single day. Interactions between the oceans and atmosphere occur slowly over many months to years, and so does the movement of water within the oceans, including the mixing of deep and shallow waters. Thus, trends can persist for decades, centuries, or longer. For this reason, even if greenhouse gas emissions were stabilized tomorrow, it would take many more years—decades to centuries—for the oceans to adjust to changes in the atmosphere and the climate that have already occurred.

Ocean Heat

This indicator describes trends in the amount of heat stored in the world's oceans.

When sunlight reaches the Earth's surface, the world's oceans absorb this energy as heat, which currents distribute around the world. Water has a much higher heat capacity than air, meaning that oceans can absorb large amounts of heat with only a slight increase in temperature. As a result, increasing concentrations of heat-trapping greenhouse gases have not caused the oceans to warm as much as the atmosphere, even though they have absorbed more than 90 percent of the Earth's extra heat since 1955.[1,2] If not for the large heat-storage capacity provided by the oceans, the atmosphere would grow warmer more rapidly.[3] Water temperature reflects the amount of heat in the water at a particular time and location, and it plays an important role in the Earth's climate system, because heat from ocean surface waters provides energy for storms, influences weather patterns, and can change ocean currents. Because water expands slightly as it gets warmer, an increase in ocean heat content will also increase the volume of water in the ocean, which is one cause of the observed increases in sea level (see the Sea Level indicator on p. 34).

WHAT'S HAPPENING

- In three different data analyses, the long-term trend shows that the oceans have become warmer since 1955.

- Although concentrations of greenhouse gases have risen at a relatively steady rate over the past few decades (see the Atmospheric Concentrations of Greenhouse Gases indicator on p. 15), the rate of change in ocean heat content can vary from year to year. Year-to-year changes are influenced by events such as volcanic eruptions and recurring ocean-atmosphere patterns such as El Niño.

Ocean Heat Content, 1955–2015

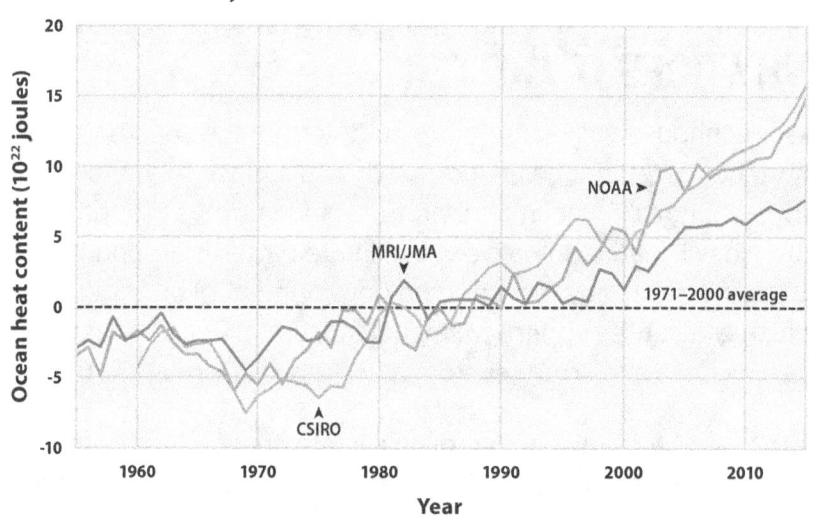

This figure shows changes in ocean heat content between 1955 and 2015. Ocean heat content is measured in joules, a unit of energy, and compared against the 1971–2000 average, which is set at zero for reference. Choosing a different baseline period would not change the shape of the data over time. The lines were independently calculated using different methods by government agencies in three countries. For reference, an increase of 1 unit on this graph (1 x 10^{22} joules) is equal to approximately 18 times the total amount of energy used by all the people on Earth in a year.[4] Data sources: CSIRO, 2016;[5] MRI/JMA, 2016;[6] NOAA, 2016[7]

ABOUT THE INDICATOR

This indicator measures changes in the amount of heat energy stored in the ocean, based on measurements of ocean temperatures around the world at different depths. These measurements come from a variety of instruments deployed from ships and airplanes and, more recently, underwater robots. Thus, the data are carefully adjusted to account for differences among measurement techniques and data collection programs. This indicator is based on analyses conducted by government agencies in three countries: the National Oceanic and Atmospheric Administration, the Japan Meteorological Agency's Meteorological Research Institute, and Australia's Commonwealth Scientific and Industrial Research Organisation.

Sea Surface Temperature

This indicator describes global trends in sea surface temperature.

As the oceans absorb more heat, sea surface temperature increases and the ocean circulation patterns that transport warm and cold water around the globe change, affecting which species are present in marine ecosystems, altering migration and breeding patterns, threatening corals, and changing the frequency and intensity of harmful algal blooms.[8] Over the long term, increases in sea surface temperature could weaken the circulation patterns that bring nutrients from the deep sea to surface waters, contributing to declines in fish populations that would affect people who depend on fishing for food or jobs.[9] Higher sea surface temperature causes an increase in the amount of atmospheric water vapor, which increases the risk of heavy rain and snow (see the Heavy Precipitation and Tropical Cyclone Activity indicators on pp. 24 and 25).[10] Changes in sea surface temperature can also shift storm tracks, potentially contributing to droughts in some areas.[11]

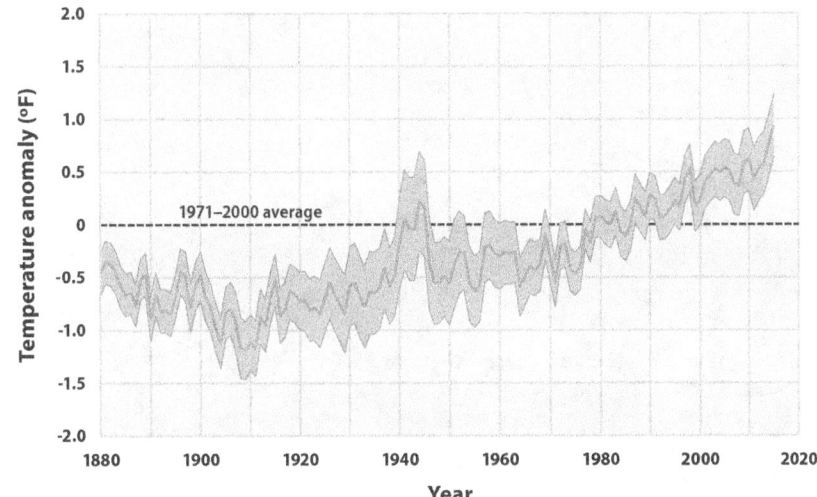

Average Global Sea Surface Temperature, 1880–2015

This graph shows how the average surface temperature of the world's oceans has changed since 1880. This graph uses the 1971 to 2000 average as a baseline for depicting change. Choosing a different baseline period would not change the shape of the data over time. The shaded band shows the range of uncertainty in the data, based on the number of measurements collected and the precision of the methods used. Data source: NOAA, 2016[12]

WHAT'S HAPPENING

- Sea surface temperature increased during the 20th century and continues to rise. From 1901 through 2015, temperature rose at an average rate of 0.13°F per decade.

- Sea surface temperature has been consistently higher during the past three decades than at any other time since reliable observations began in 1880.

- Based on the historical record, increases in sea surface temperature have largely occurred over two key periods: between 1910 and 1940, and from about 1970 to the present. Sea surface temperature appears to have cooled between 1880 and 1910.

Oceans

HEALTH CONNECTION

Rising sea surface temperature means that *Vibrio* bacteria and blooms of harmful algae can occur in new places or at new times of the year. Humans can be exposed to *Vibrio* and algal toxins by eating contaminated seafood or through direct contact with contaminated drinking or recreational waters. *Vibrio* can cause gastrointestinal illness and bloodstream infections; algal toxins can cause gastrointestinal illness and neurologic symptoms; and both can cause death in severe cases.[13]

ABOUT THE INDICATOR

The global average sea surface temperature data shown here are derived from the Extended Reconstructed Sea Surface Temperature analysis developed by the National Oceanic and Atmospheric Administration's (NOAA's) National Centers for Environmental Information. Temperature measurements are collected from ships, as well as at stationary and drifting buoys. NOAA has carefully reconstructed and filtered the data in the figure to correct for biases in different collection techniques and to minimize the effects of sampling changes over various locations and times. The data are shown as anomalies, or differences, compared with the average sea surface temperature from 1971 to 2000. The online version of this indicator also presents global changes in sea surface temperature in a map.

Sea Level

This indicator describes how sea level has changed over time. The indicator describes two types of sea level changes: absolute and relative.

A s the temperature of the Earth changes, so does sea level. Temperature and sea level are linked for two main reasons. First, changes in the volume of water and ice on land (namely glaciers and ice sheets) can increase or decrease the volume of water in the ocean (see the Glaciers indicator on p. 44). Second, as water warms, it expands slightly—an effect that is cumulative over the entire depth of the oceans (see the Ocean Heat indicator on p. 32). Rising sea level inundates low-lying wetlands and dry land, erodes shorelines, contributes to coastal flooding, and increases the flow of salt water into estuaries and nearby groundwater aquifers. Higher sea level also makes coastal infrastructure more vulnerable to damage from storms, due to an increased likelihood of flooding from higher storm surges.

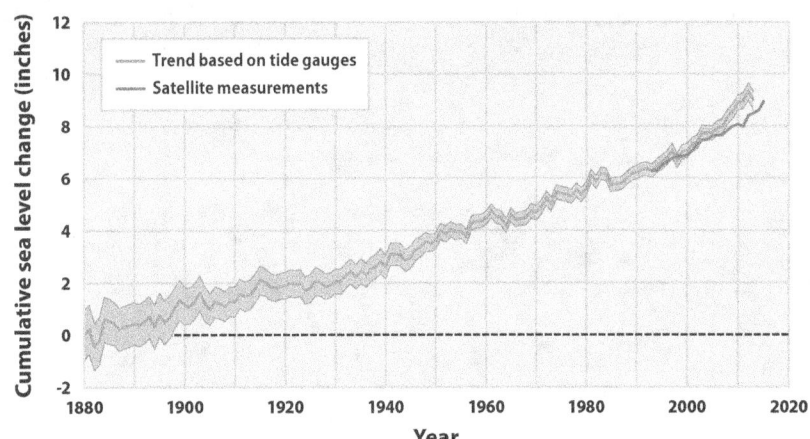

Global Average Absolute Sea Level Change, 1880–2015

- Trend based on tide gauges
- Satellite measurements

This graph shows cumulative changes in sea level for the world's oceans since 1880, based on a combination of long-term tide gauge measurements and recent satellite measurements. This figure shows average absolute sea level change, which refers to the height of the ocean surface, regardless of whether nearby land is rising or falling. Satellite data are based solely on measured sea level, while the long-term tide gauge data include a small correction factor because the size and shape of the oceans are changing slowly over time. (On average, the ocean floor has been gradually sinking since the last Ice Age peak, 20,000 years ago.) The shaded band shows the likely range of values, based on the number of measurements collected and the precision of the methods used. Data sources: CSIRO, 2015;[14] NOAA, 2016[15]

WHAT'S HAPPENING

- After a period of approximately 2,000 years of little change (not shown here), global average sea level rose throughout the 20th century, and the rate of change has accelerated in recent years.[16] When averaged over all of the world's oceans, absolute sea level has risen at an average rate of 0.06 inches per year from 1880 to 2013. Since 1993, however, average sea level has risen at a rate of 0.11 to 0.14 inches per year—roughly twice as fast as the long-term trend.

ABOUT THE INDICATOR

Scientists measure sea level change in two different ways. *Relative* sea level change refers to how the height of the ocean rises or falls relative to the land at a particular location. In contrast, *absolute* sea level change refers to the height of the ocean surface above the center of the Earth, without regard to nearby land. This distinction matters because the land itself can rise or fall relative to the ocean—rising due to processes such as sediment accumulation and geological uplift, or falling because of erosion, sediment compaction, natural subsidence, groundwater withdrawal, or engineering projects that prevent rivers from naturally depositing sediments along their banks. The first graph above shows absolute sea level change averaged across the Earth's oceans since 1880. The long-term trend is based on tide gauges, which measure relative sea level change but have been adjusted to show absolute trends through calibration with recent satellite data. The second graph shows the change in relative sea level based on tide gauges that have measured water levels at 67 points along the U.S. coast since at least 1960. The National Oceanic and Atmospheric Administration and Australia's Commonwealth Scientific and Industrial Research Organisation compiled the data for this indicator.

- Relative sea level rose along much of the U.S. coastline between 1960 and 2015, particularly the Mid-Atlantic coast and parts of the Gulf coast, where some stations registered increases of more than 8 inches. Meanwhile, relative sea level fell at some locations in Alaska and the Pacific Northwest. At those sites, even though absolute sea level has risen, land elevation has risen more rapidly.

- Relative sea level also has not risen uniformly because of regional and local changes in land movement and long-term changes in coastal circulation patterns.

This map shows cumulative changes in relative sea level from 1960 to 2015 at tide gauge stations along U.S. coasts. Relative sea level reflects changes in sea level as well as land elevation. Data source: NOAA, 2016[17]

Relative Sea Level Change Along U.S. Coasts, 1960–2015

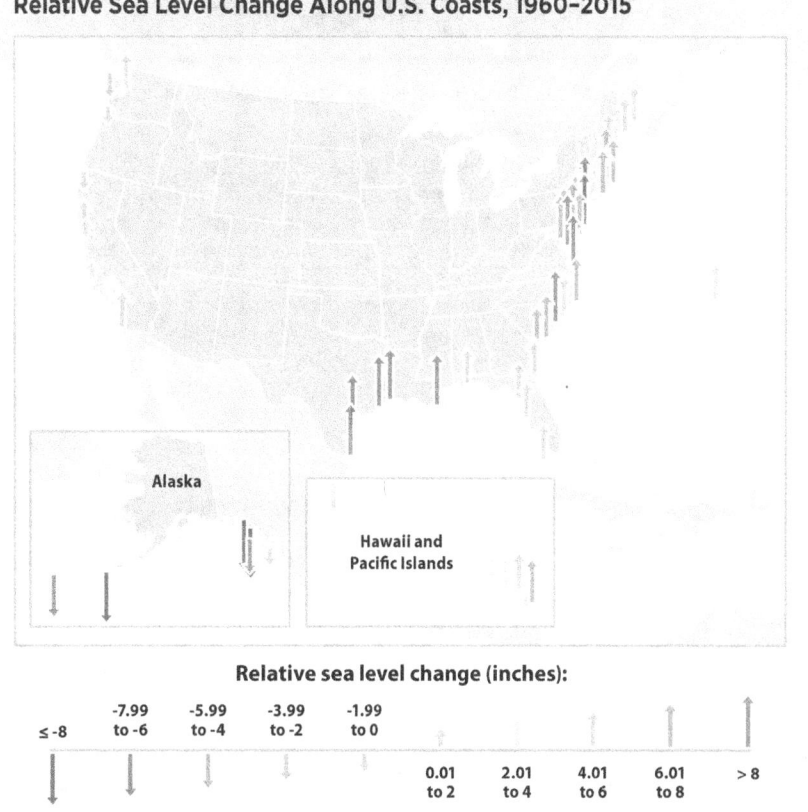

Relative sea level change (inches):

| ≤ -8 | -7.99 to -6 | -5.99 to -4 | -3.99 to -2 | -1.99 to 0 | 0.01 to 2 | 2.01 to 4 | 4.01 to 6 | 6.01 to 8 | > 8 |

Oceans

A CLOSER LOOK: LAND LOSS ALONG THE ATLANTIC COAST

The Atlantic coast is particularly vulnerable to rising sea level because of its low elevations and sinking shorelines. The graph below shows the amount of land lost to sea level rise along the Atlantic coast from Florida to New York, dividing the Atlantic coast into two regions for purposes of comparison. It is based on satellite imagery from the National Oceanic and Atmospheric Administration's Coastal Change Analysis Program. These data have been collected and analyzed at five-year intervals since 1996. Roughly 20 square miles of dry land and wetlands were converted to open water along the Atlantic coast between 1996 and 2011. More of this loss occurred in the Southeast than in the Mid-Atlantic.

Land Loss Along the Atlantic Coast, 1996–2011

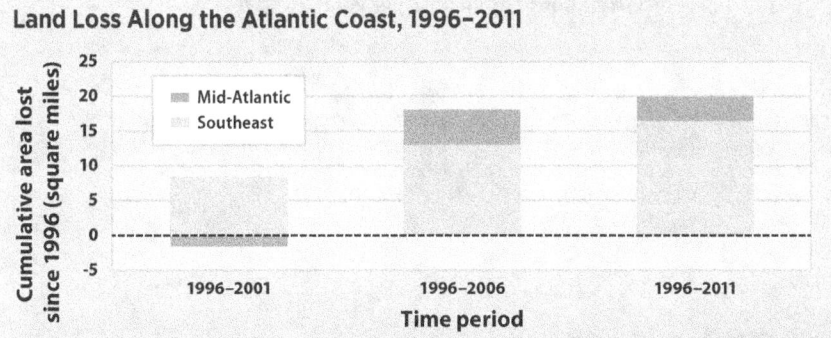

This graph shows the net amount of land converted to open water along the Atlantic coast during three time periods: 1996–2001, 1996–2006, and 1996–2011. The results are divided into two regions: the Southeast and the Mid-Atlantic. Negative numbers show where land loss is outpaced by the accumulation of new land. Data source: NOAA, 2013[18]

Coastal Flooding

This indicator shows how the frequency of coastal flooding has changed over time.

As sea level rises relative to the coast due to climate change (see the Sea Level indicator on p. 34), one of the most noticeable consequences is an increase in coastal flooding during high tide and during storm surges. Many coastal cities have defined minor or "nuisance" flooding thresholds. When water rises above this level, minor flooding typically occurs in some streets, many storm drains become ineffective, and a coastal flood advisory may be issued. Recurrent coastal flooding can cause impacts such as frequent road closures, reduced stormwater drainage capacity, and deterioration of infrastructure not designed to withstand frequent inundation or exposure to salt water. Rising sea level increases the likelihood of flooding at high tide and during storm surges, and it also inundates low-lying wetlands and dry land, erodes shorelines, and increases the flow of salt water into estuaries and nearby groundwater aquifers. Millions of Americans and more than $1 trillion of property and infrastructure are at risk of damage from coastal flooding.[19]

WHAT'S HAPPENING

- Flooding is becoming more frequent along the U.S. coastline. Nearly every site measured has experienced an increase in coastal flooding since the 1950s. The rate is accelerating in many locations along the East and Gulf coasts.

- The Mid-Atlantic region suffers the highest number of coastal flood days and has also experienced the largest increases in flooding. Since 2010, Wilmington, North Carolina, has flooded most often—49 days per year—followed by Annapolis, Maryland, at 46 days per year. Annapolis, Wilmington, and two locations in New Jersey (Sandy Hook and Atlantic City) have also seen some of the most dramatic overall increases in frequency: floods are now at least 10 times more common there than they were in the 1950s. The Mid-Atlantic's subsiding land and higher-than-average relative sea level rise both contribute to this increase in flooding (see the Sea Level indicator on p. 34).

Frequency of Flooding Along U.S. Coasts, 2010–2015 Versus 1950–1959

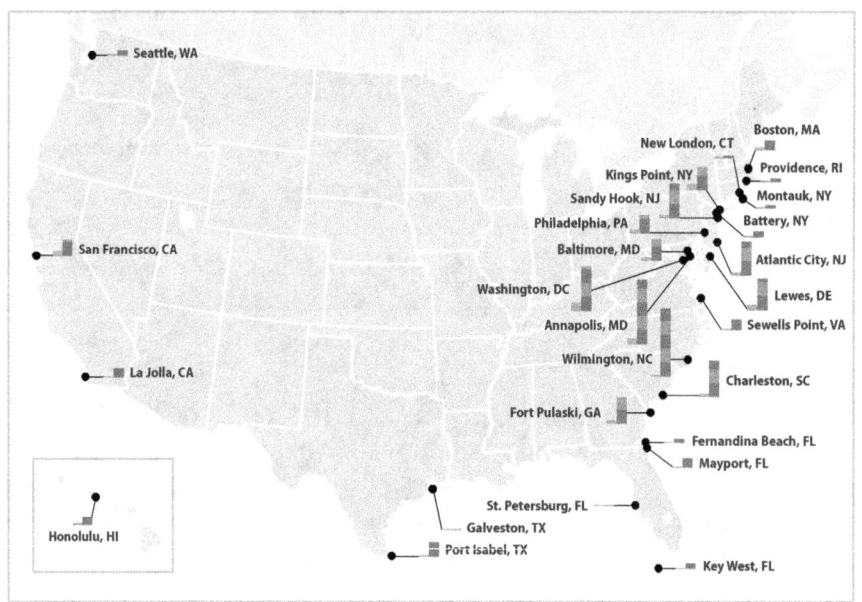

Average number of flood days per year:

This map shows the average number of days per year in which coastal waters rose above the local threshold for minor flooding at 27 sites along U.S. coasts. Each small bar graph compares the first decade of widespread measurements (the 1950s in orange) with the most recent decade (the 2010s in purple). Data source: NOAA, 2016[20]

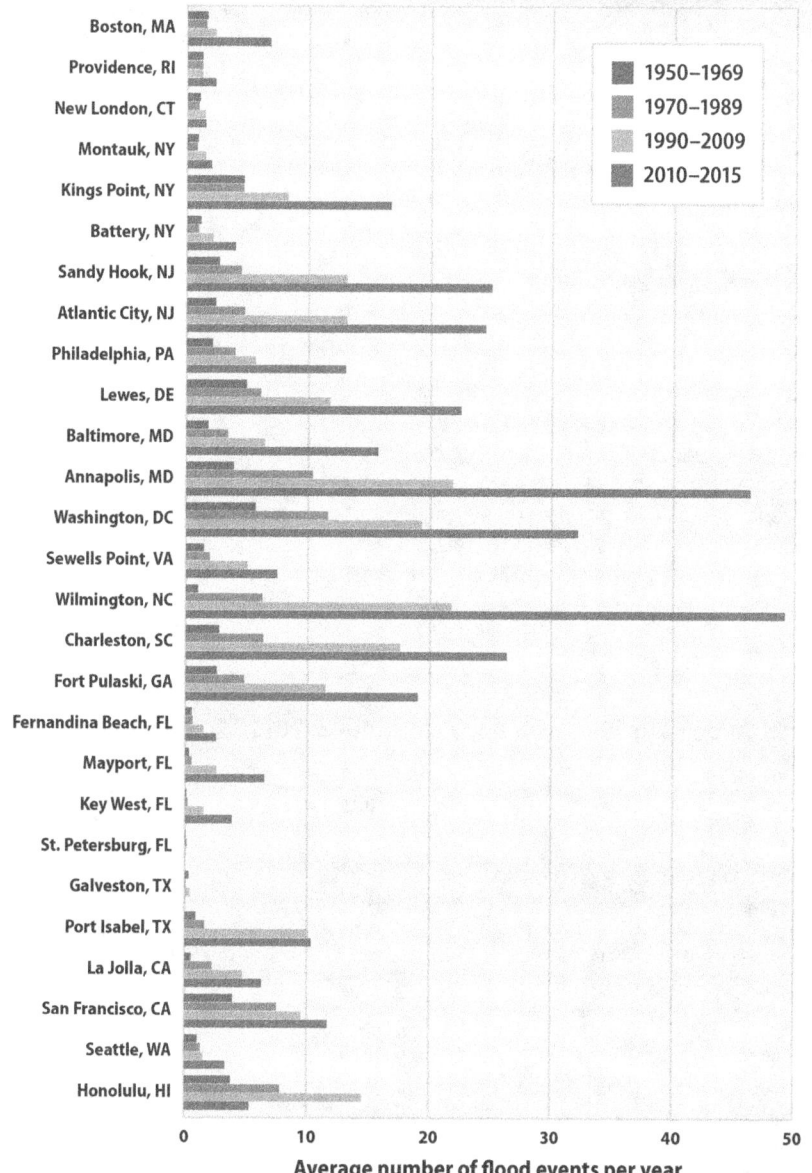

Average Number of Coastal Flood Events per Year, 1950–2015

Legend:
- 1950–1969
- 1970–1989
- 1990–2009
- 2010–2015

(Cities, top to bottom: Boston, MA; Providence, RI; New London, CT; Montauk, NY; Kings Point, NY; Battery, NY; Sandy Hook, NJ; Atlantic City, NJ; Philadelphia, PA; Lewes, DE; Baltimore, MD; Annapolis, MD; Washington, DC; Sewells Point, VA; Wilmington, NC; Charleston, SC; Fort Pulaski, GA; Fernandina Beach, FL; Mayport, FL; Key West, FL; St. Petersburg, FL; Galveston, TX; Port Isabel, TX; La Jolla, CA; San Francisco, CA; Seattle, WA; Honolulu, HI)

X-axis: 0, 10, 20, 30, 40, 50
Average number of flood events per year

- Flooding has increased less dramatically in places where the local flood threshold is higher (for example, the Northeast and locations on the Gulf of Mexico) or where relative sea level has not risen as quickly as it has elsewhere in the United States (for example, Hawaii and the West Coast, as shown by the Sea Level indicator on p. 34).

This graph shows the average number of days per year in which coastal waters rose above the local threshold for minor flooding at 27 sites along U.S. coasts. The data have been averaged over multi-year periods for comparison.
Data source: NOAA, 2016[21]

Oceans

ABOUT THE INDICATOR

Coastal flooding trends in this indicator are based on measurements from 27 permanent tide gauge stations along U.S. coasts where local weather forecasting offices have defined thresholds for minor, moderate, and major flooding and where complete data are available from 1950 to present. The indicator tracks the number of days per year when each tide gauge measured water that was higher than the minor flooding level. The original tide gauge data and the analysis come from the National Oceanic and Atmospheric Administration, which derived daily maximum water levels from hourly data.

HEALTH CONNECTION

Recurrent coastal flooding can increase the risk that drinking water and wastewater infrastructure will fail, putting people at risk of exposure to pathogens and harmful chemicals.[22] Heavy rain during high tides can lead to flooding of basements and standing water in streets, which can also harbor disease-carrying vectors such as mosquitoes. Extreme flood events that involve loss of life or homes are also associated with long-term anxiety, depression, and post-traumatic stress disorder.[23]

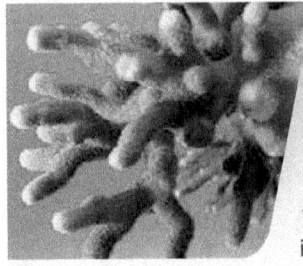

Ocean Acidity

This indicator describes changes in the chemistry of the ocean that relate to the amount of carbon dioxide dissolved in the water.

As the concentration of carbon dioxide in the atmosphere increases, the ocean absorbs more of it. Over the past 250 years, oceans have absorbed about 28 percent of the carbon dioxide produced by human activities that burn fossil fuels.[24] Rising levels of carbon dioxide dissolved in the ocean negatively affect some marine life, because carbon dioxide reacts with sea water to produce carbonic acid. The increase in acidity changes the balance of minerals in the water and makes it more difficult for corals and plankton to produce the mineral calcium carbonate, which is the primary component of their hard skeletons and shells. Resulting declines in coral and plankton populations can change marine ecosystems and ultimately affect fish populations and the people who depend on them.[25] Signs of damage are already starting to appear in certain areas.[26]

ABOUT THE INDICATOR

This indicator describes trends in pH and related properties of ocean water, based on a combination of direct observations, calculations, and modeling. The graph shows pH values and levels of dissolved carbon dioxide at three locations that have collected measurements consistently over the last few decades. These data have been either measured directly or calculated from related measurements, such as dissolved inorganic carbon and alkalinity. Data come from two stations in the Atlantic Ocean (Bermuda and the Canary Islands) and one in the Pacific (Hawaii). The online version of this indicator shows a map of changes in aragonite saturation of the world's oceans. Aragonite is a form of calcium carbonate that many organisms produce and use to build their protective skeletons or shells. Saturation state is a measure of how easily aragonite can dissolve in the water.

WHAT'S HAPPENING

- Measurements made over the last few decades have demonstrated that ocean carbon dioxide levels have risen in response to increased carbon dioxide in the atmosphere, leading to an increase in acidity (that is, a decrease in pH).

Ocean Carbon Dioxide Levels and Acidity, 1983–2015

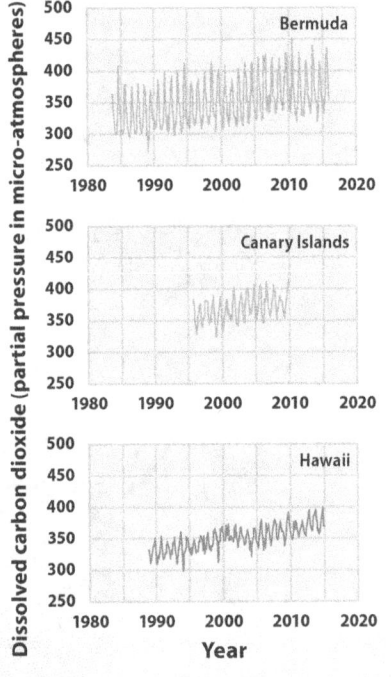

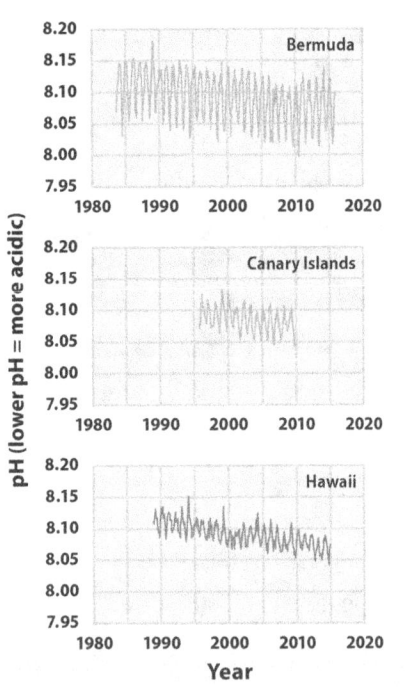

This figure shows the relationship between changes in ocean carbon dioxide levels (measured in the left column as a partial pressure—a common way of measuring the amount of a gas) and acidity (measured as pH in the right column). The data come from three observation stations. The up-and-down pattern shows the influence of seasonal variations.
Data sources: Bates, 2016;[27] González-Dávila, 2012;[28] Dore, 2015[29]

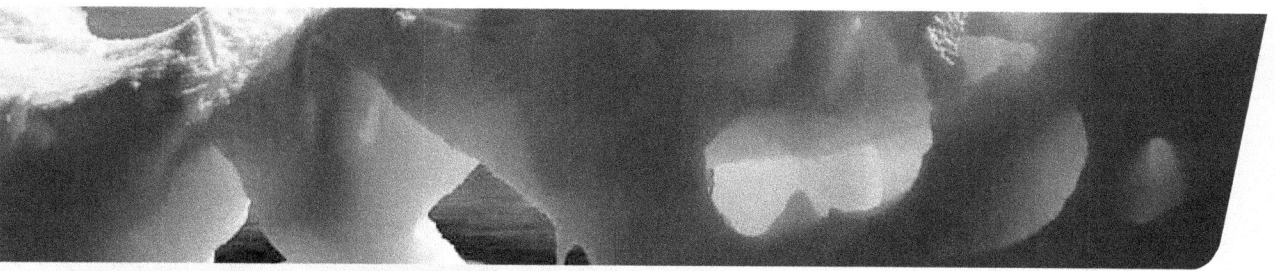

 # Snow and Ice

The Earth's surface contains many forms of snow and ice, including sea, lake, and river ice; snow cover; glaciers, ice caps, and ice sheets; and frozen ground. Climate change can dramatically alter the Earth's snow- and ice-covered areas because snow and ice can easily change between solid and liquid states in response to relatively minor changes in temperature. This chapter focuses on trends in snow, glaciers, and the freezing and thawing of oceans and lakes.

WHY DOES IT MATTER?

Reduced snowfall and less snow cover on the ground could diminish the beneficial insulating effects of snow for vegetation and wildlife, while also affecting water supplies, transportation, cultural practices, travel, and recreation for millions of people. For communities in Arctic regions, reduced sea ice could increase coastal erosion and exposure to storms, threatening homes and property, while thawing ground could damage roads and buildings and accelerate erosion. Conversely, reduced snow and ice could present commercial opportunities for others, including ice-free shipping lanes and increased access to natural resources.

Such changing climate conditions can have worldwide implications because snow and ice influence air temperatures, sea level, ocean currents, and storm patterns. For example, melting ice sheets on Greenland and Antarctica add fresh water to the ocean, increasing sea level and possibly changing ocean circulation that is driven by differences in temperature and salinity. Because of their light color, snow and ice also reflect more sunlight than open water or bare ground, so a reduction in snow cover and ice causes the Earth's surface to absorb more energy from the sun and become warmer.

Arctic Sea Ice

This indicator tracks the extent, age, and melt season of sea ice in the Arctic Ocean.

Sea ice is an integral part of the Arctic Ocean. Each year some of this ice melts during the summer because of warmer temperatures and sunlight, typically reaching its minimum thickness and extent in mid-September. The ice freezes and begins expanding again in the fall. Sea ice extent is an important indicator of global climate change because warmer air and water temperatures are reducing the amount of sea ice present. Sea ice reflects sunlight, which helps to keep polar regions cool. Sea ice is also important because it provides habitat for animals such as polar bears and walruses, and because wildlife and ice travel are vital to the traditional subsistence lifestyle of indigenous Arctic communities.

WHAT'S HAPPENING

- September 2012 had the lowest sea ice extent ever recorded, 44 percent below the 1981–2010 average for that month.

- The September 2015 sea ice extent was more than 700,000 square miles less than the historical 1981–2010 average for that month—a difference more than two and a half times the size of Texas. March sea ice extent reached the lowest extent on record in 2015 and hit roughly the same low again in 2016— about 7 percent less than the 1981–2010 average.

Dwindling Arctic Sea Ice

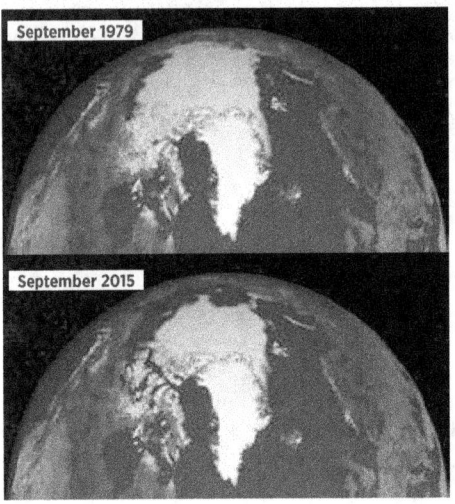

Source: NASA, 2016[1]

March and September Monthly Average Arctic Sea Ice Extent, 1979–2016

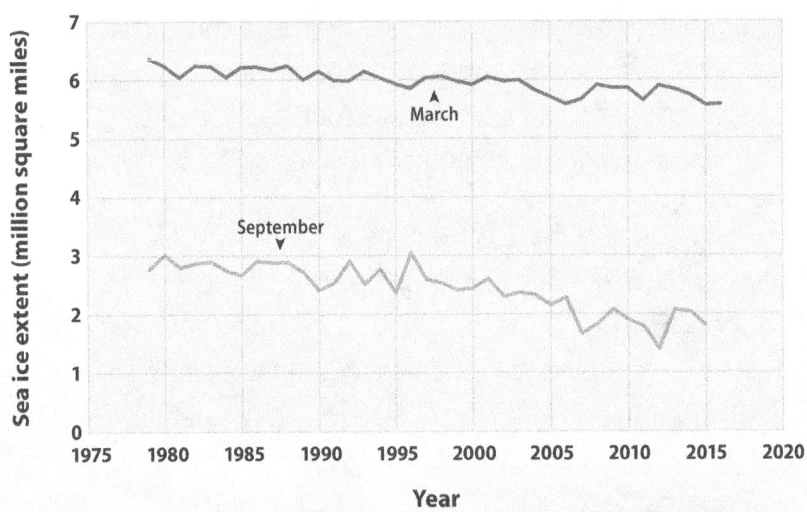

This figure shows Arctic sea ice extent for the months of September and March of each year from 1979 through September 2015 and March 2016. September and March are when the minimum and maximum extent typically occur each year. Data source: NSIDC, 2016[2]

Age of Arctic Sea Ice at Minimum September Week, 1983–2015

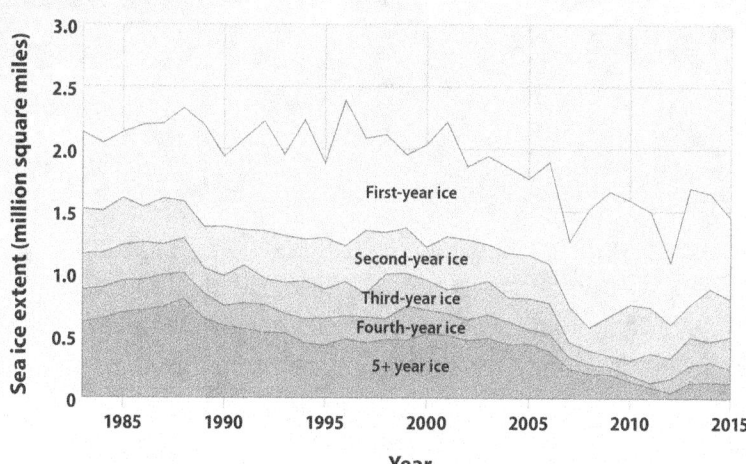

This figure shows the distribution of Arctic sea ice extent by age group during the week in September with the smallest extent of ice for each year. Total extent differs between this figure and the one on p. 40 that one shows a monthly average, while this one shows conditions during a single week. Data source: NSIDC, 2015[3]

Arctic Sea Ice Melt Season, 1979–2015

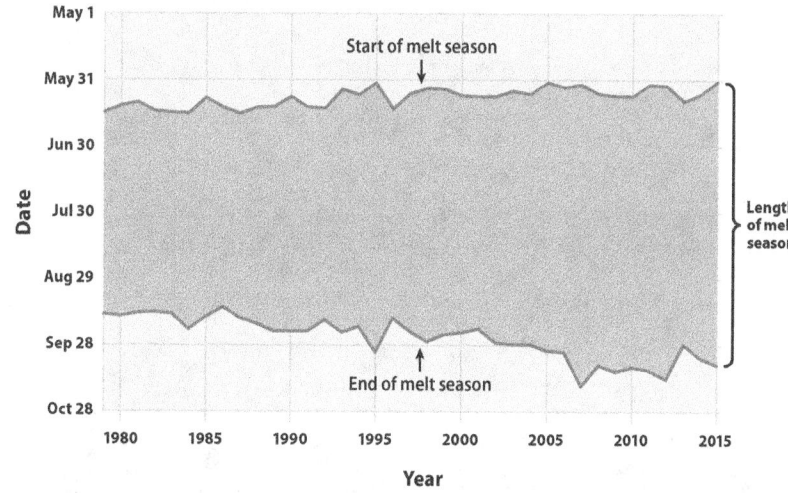

This figure shows the timing of each year's Arctic sea ice melt season. The shaded band spans from the date when ice begins to melt consistently until the date when it begins to refreeze. Data source: NASA, 2016[4]

- Evidence of the age of Arctic sea ice suggests that fewer patches of ice are persisting for multiple years (i.e., generally thick ice that has survived one or more melt seasons). The proportion of sea ice five years or older has declined dramatically over the recorded time period, from more than 30 percent of September ice in the 1980s to 9 percent in 2015. A growing percentage of Arctic sea ice is only one or two years old. Less old multi-year ice implies that the ice cover is thinning, which makes it more vulnerable to further melting.

- Since 1979, the length of the melt season for Arctic sea ice has grown by 37 days. Arctic sea ice now starts melting 11 days earlier and it starts refreezing 26 days later than it used to, on average.

Snow & Ice

ABOUT THE INDICATOR

This indicator shows the extent of sea ice in the Arctic region, which is defined as the area of ocean where at least 15 percent of the surface is frozen. It also examines the age distribution of sea ice and the start and end dates of each year's Arctic sea ice melt season. This indicator is based on routine monitoring of sea ice conditions from satellite measurements, which began in 1979. Here, the melt season start date is defined as the date when satellites detect consistent wetness on the surface of the ice and snow; the end date is when the surface air temperature stays consistently at or below the freezing point and ice begins to grow in the open ocean. Data for this indicator were gathered by the National Snow and Ice Data Center using satellite imaging technology and data processing methods developed by the National Aeronautics and Space Administration and the University of Colorado, Boulder.

Collecting
SNOW AND ICE DATA

From "low-tech" backyard observations made by citizens in their own neighborhood to "high-tech" global satellite images of some of the most remote places in the world, scientists use a variety of techniques to track climate change. Below are some examples of the data collection methods used to create the indicators in this chapter.

SATELLITES

Satellites provide an efficient way to collect the kind of data that would be difficult to measure in person, such as measurements that need to be made at regular intervals over large areas or in remote locations. For example, the Arctic Sea Ice indicator (p. 40) is derived from data-rich images taken by satellites that orbit the Earth every day, using instruments that can tell the difference between sea ice and open water.

FIELD MEASUREMENTS

Sometimes going out in the field and measuring by hand is the best way to collect precise information and maintain a long-running dataset. For the Glaciers indicator (p. 44), scientists visit the same glaciers twice a year at locations marked with a network of stakes, where they measure snow depth and density.

CITIZEN SCIENCE

You don't need a Ph.D. to be a scientist. Many citizen scientists have helped to create high-quality datasets of climate indicators that date back further than some modern climate monitoring programs. The Lake Ice indicator (p. 46) includes local observations made from the same vantage point throughout the year. Some lakes have multiple observers, such as residents on both sides of a lake who can compare notes.

OBSERVATION STATIONS

Automated observation stations make it possible to collect data continuously from places that may be difficult or expensive to reach. For example, scientists once had to travel to remote snowpack measurement sites by ski, snowshoe, snowmobile, or helicopter. Now, the Snowpack indicator (p. 52) uses hundreds of snow telemetry stations that automatically record and relay data back to a central computer every 15 minutes.

Antarctic Sea Ice

This indicator tracks the extent of sea ice around Antarctica.

The Southern Ocean around Antarctica freezes to form sea ice every year. This sea ice reaches its maximum extent in September or early October and melts in the summer months (December to February). Like Arctic sea ice (see the Arctic Sea Ice indicator on p. 40), Antarctic sea ice affects global climate, regional climate, and ecosystems. Unlike the Arctic, where a large area of sea ice lasts year-round, the sea ice around Antarctica is thinner, and nearly all of it melts in a typical summer. Warmer air and ocean temperatures are generally expected to reduce the amount of sea ice present worldwide. While warming has already driven a noticeable decline in sea ice in the Arctic, extent for the Antarctic as a whole has not declined (and has actually increased slightly), which may reflect influences of wind patterns, ocean currents, and precipitation around the continent.[5]

February and September Monthly Average Antarctic Sea Ice Extent, 1979–2016

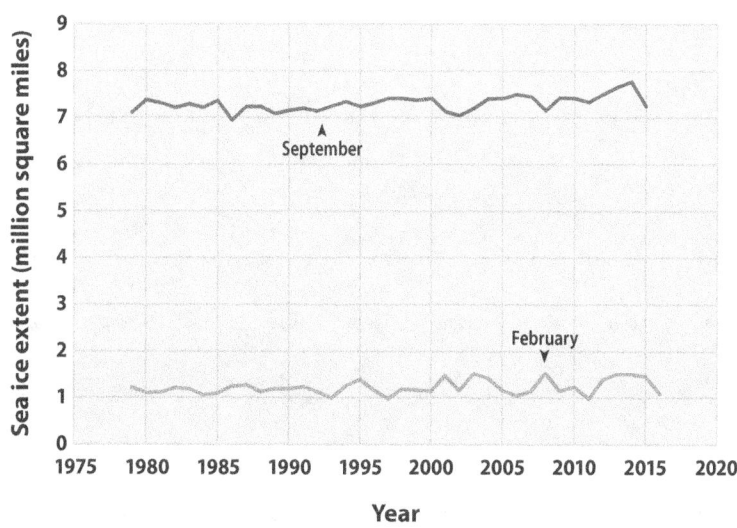

This figure shows Antarctic sea ice extent for the months of February and September of each year from 1979 through September 2015 and February 2016. February and September typically have the minimum and maximum extent each year. Data source: NSIDC, 2016[6]

WHAT'S HAPPENING

- Antarctic sea ice extent in September and February has increased somewhat over time, although the most recent year was below average. The September maximum extent reached the highest level on record in 2014—about 7 percent larger than the 1981–2010 average—but in 2015 it was slightly below the 1981–2010 average. As for February extent, 2013, 2014, and 2015 were three of the six largest years on record, but extent in 2016 was about 9 percent below the 1981–2010 average.

- Slight increases in Antarctic sea ice are outweighed by the loss of sea ice in the Arctic during the same time period (see the Arctic Sea Ice indicator on p. 40). Overall, the Earth has lost sea ice at an average rate of 13,500 square miles per year since 1979—equivalent to losing an area larger than the state of Maryland every year.[7] This decrease affects the Earth's energy balance.

Snow & Ice

ABOUT THE INDICATOR

This indicator examines the extent of sea ice in the Southern Ocean, which is defined as the area of ocean where at least 15 percent of the surface is frozen. It is based on routine monitoring of sea ice conditions from satellite measurements, which began in 1979. Monthly average sea ice extent data for this indicator were gathered by the National Snow and Ice Data Center using satellite imaging technology and data processing methods developed by the National Aeronautics and Space Administration. Data are collected throughout the year, but for comparison, this indicator focuses on the months when sea ice typically reaches its minimum and maximum extent.

Glaciers

This indicator examines the balance between snow accumulation and melting in glaciers, and it describes how glaciers in the United States and around the world have changed over time.

A glacier is a large mass of snow and ice that has accumulated over many years and is present year-round. In many areas, glaciers provide communities and ecosystems with a reliable source of streamflow and drinking water, particularly in times of extended drought and late in the summer, when seasonal snowpack has melted away. Glaciers are important as an indicator of climate change because physical changes in glaciers—whether they are growing or shrinking, advancing or receding—provide visible evidence of changes in temperature and precipitation. If glaciers lose more ice than they can accumulate through new snowfall, they ultimately add more water to the oceans, leading to a rise in sea level (see the Sea Level indicator on p. 34).

WHAT'S HAPPENING

- On average, glaciers worldwide have been losing mass since at least the 1970s, which in turn has contributed to observed changes in sea level (see the Sea Level indicator on p. 34). A longer measurement record from a smaller number of glaciers suggests that they have been shrinking since the 1940s. The rate at which glaciers are losing mass appears to have accelerated over roughly the last decade.

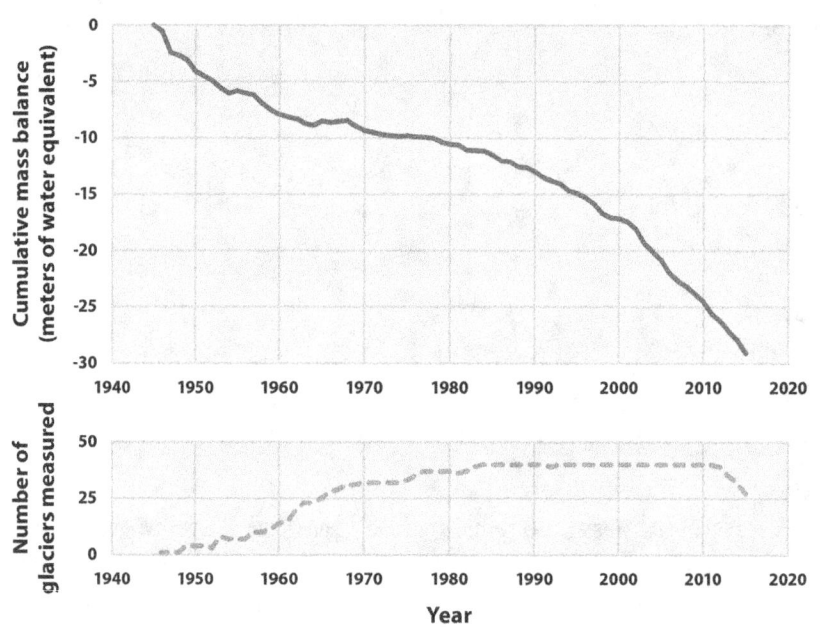

Average Cumulative Mass Balance of "Reference" Glaciers Worldwide, 1945–2015

This figure shows the cumulative change in mass balance of a set of "reference" glaciers worldwide beginning in 1945. The line on the upper graph represents the average of all the glaciers that were measured. Negative values indicate a net loss of ice and snow compared with the base year of 1945. For consistency, measurements are in meters of water equivalent, which represent changes in the average thickness of a glacier. The small chart below shows how many glaciers were measured in each year. Some glacier measurements have not yet been finalized for the last few years, hence the smaller number of sites.
Data source: WGMS, 2016[6]

Cumulative Mass Balance of Three U.S. Glaciers, 1958–2014

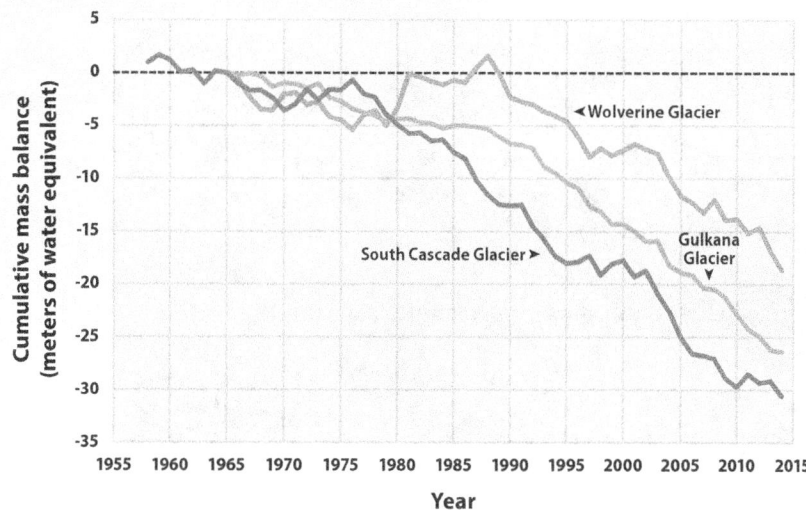

This figure shows the cumulative mass balance of the three U.S. Geological Survey "benchmark" glaciers since measurements began in the 1950s or 1960s. For each glacier, the mass balance is set at zero for the base year of 1965. Negative values indicate a net loss of ice and snow compared with the base year. For consistency, measurements are in meters of water equivalent, which represent changes in the average thickness of a glacier. Data sources: O'Neel et al., 2014;[9] USGS, 2015[10]

- All three U.S. benchmark glaciers have shown an overall decline in mass balance since the 1950s and 1960s and an accelerated rate of decline in recent years. Year-to-year trends vary, with some glaciers gaining mass in certain years (for example, Wolverine Glacier during the 1980s), but the measurements clearly indicate a loss of glacier mass over time.

- Trends for the three benchmark glaciers are consistent with the retreat of glaciers observed throughout the western United States, Alaska, and other parts of the world.[11]

ABOUT THE INDICATOR

This indicator examines changes in glacier mass balance, which is the net gain or loss of snow and ice over the course of the year. It can also be thought of as the average change in thickness across the surface of a glacier. The change in ice or snow has been converted to an equivalent amount of liquid water. If cumulative mass balance becomes more negative over time, it means glaciers are losing mass more quickly than they can accumulate new snow. The first graph above shows the average change across 40 reference glaciers around the world that have been measured consistently for many decades. The World Glacier Monitoring Service compiled these data, based on measurements collected by a variety of organizations around the world. Data for the second graph come from the U.S. Geological Survey Benchmark Glacier Program, which has studied three U.S. "benchmark" glaciers extensively for many years. These three glaciers are thought to be representative of other glaciers nearby.

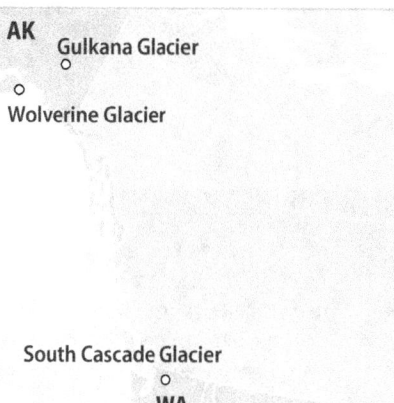

Snow & Ice

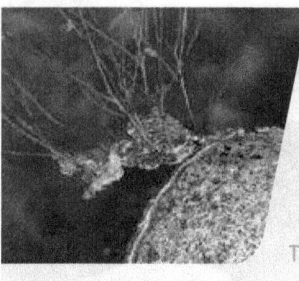

Lake Ice

This indicator measures the amount of time that ice is present on lakes in the United States.

Lake ice formation and breakup dates are key indicators of climate change. If lakes remain frozen for longer periods, it can signify that the climate is cooling. Conversely, shorter periods of ice cover suggest a warming climate. Changes in ice cover can affect the physical, chemical, and biological characteristics of a body of water. Reduced ice cover leads to increased evaporation and lower water levels, as well as an increase in water temperature and sunlight penetration, which in turn can affect plants and animals. The timing and duration of ice cover on lakes and other bodies of water can also affect society—particularly in relation to shipping and transportation, hydroelectric power generation, and fishing.

WHAT'S HAPPENING

• All of the lakes shown here were found to be thawing earlier in the year. Spring thaw dates have grown earlier by up to 24 days in the past 110 years.

Change in Ice Thaw Dates for Selected U.S. Lakes, 1905–2015

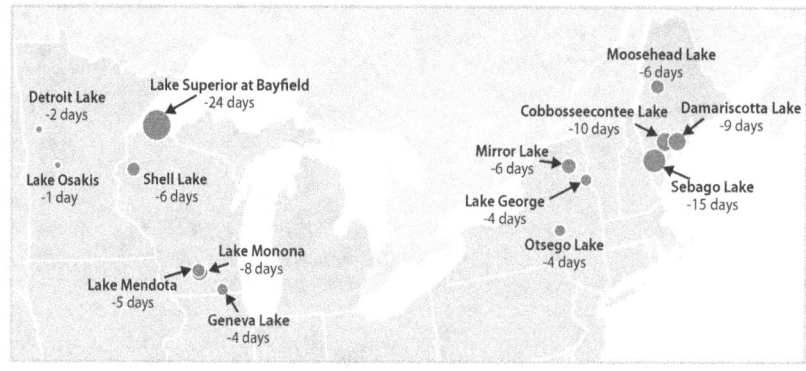

Change in ice thaw date:

● Earlier

This figure shows the change in the "ice-off" date, or date of ice thawing and breakup, for 14 U.S. lakes during the period from 1905 to 2015. All of the lakes have red circles with negative numbers, which represent earlier thaw dates. Larger circles indicate larger changes. Data source: Various organizations[12]

ABOUT THE INDICATOR

This indicator analyzes thaw dates, which occur when the ice cover on a lake breaks up, and open water becomes extensive. Thaw dates have been recorded through human visual observations for more than 100 years. Historical records from many lakes have been compiled in the Global Lake and River Ice Phenology Database, which was developed by the University of Wisconsin–Madison. Data for other lakes have been maintained by local officials or published in local newspapers. This indicator focuses on 14 lakes within the United States that have the longest and most complete historical records. The online version of this indicator tracks thaw dates for a smaller set of lakes dating back to 1840, and it also tracks changes in freeze dates, which occur when a continuous and immobile ice cover forms on the lake. Freeze dates come from visual observations recorded by the same sources as the thaw dates shown here.

COMMUNITY CONNECTION:
ICE BREAKUP IN TWO ALASKAN RIVERS

The Tanana and Yukon rivers in Alaska provide a particularly noteworthy record of northern climate because, for a century or more, local citizens have recorded the date when the ice on these rivers starts to move or break up each spring. In fact, the towns of Nenana, Alaska, and Dawson City, just over the border in Canada, hold annual competitions to guess when ice breakup will occur. To measure the exact time of ice breakup, residents place a tripod on the ice in the center of the river. This tripod is attached by a cable to a clock on the shore, so that when the ice under the tripod breaks or starts to move, the tripod will move and pull the cable, stopping the clock with the exact date and time of the river ice breakup. The Tanana and Yukon rivers both demonstrate long-term trends toward earlier ice breakup in the spring. Ice breakup dates for both rivers have shifted earlier by approximately seven days over their respective periods of record, and 2016 had the earliest breakup on record at Dawson City. However, other recent breakup dates for both rivers are within the range of historical variation.

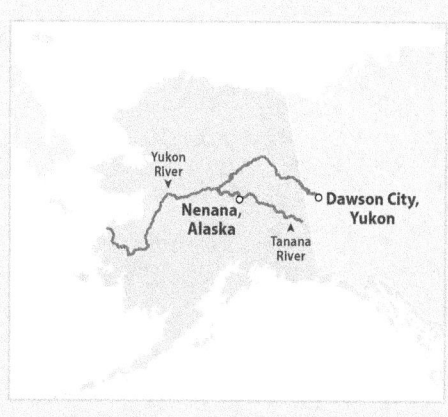

Ice Breakup Dates for Two Alaskan Rivers, 1896–2016

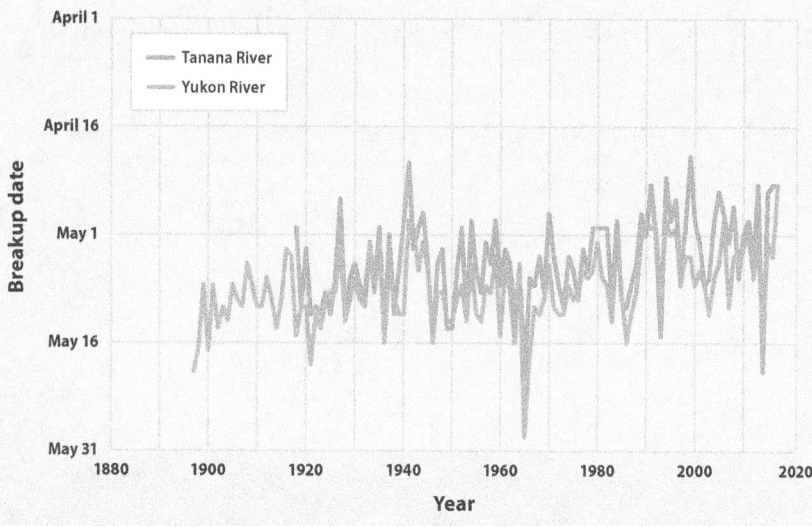

This figure shows the date each year when ice breaks up at two locations: the town of Nenana on the Tanana River and Dawson City on the Yukon River. Data sources: Nenana Ice Classic, 2016;[13] Yukon River Breakup, 2016[14]

Snowfall

This indicator uses two different measures to show how snowfall has changed in the contiguous 48 states.

Snowfall is an important aspect of winter in much of the United States. Warmer temperatures cause more water to evaporate from the land and oceans, which leads to more precipitation, larger storms, and more variation in precipitation in some areas. In general, a warmer climate will cause more of this precipitation to fall in the form of rain instead of snow. Some places could see more snowfall, however, if temperatures rise but still remain below the freezing point, or if storm tracks change. Changes in the amount and timing of snowfall could affect the spawning of fish in the spring and the amount of water available for people to use in the spring and summer. Changes in snowfall could also affect winter recreation activities, like skiing, and communities that rely on these activities.

WHAT'S HAPPENING

- Total snowfall has decreased in many parts of the country since widespread observations became available in 1930, with 57 percent of stations showing a decline. Among all of the stations shown, the average change is a decrease of 0.19 percent per year.

Change in Total Snowfall in the Contiguous 48 States, 1930–2007

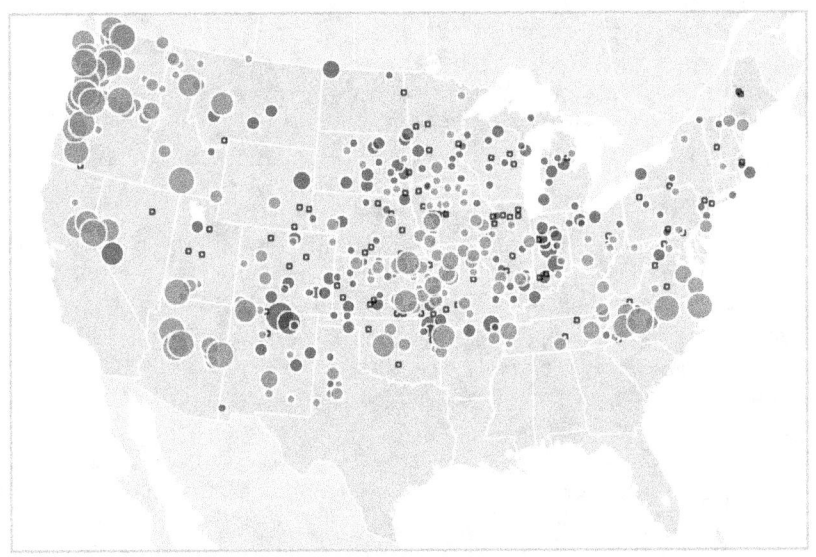

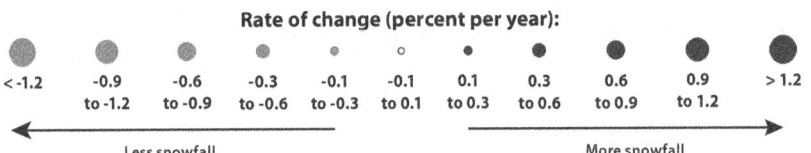

This figure shows the average rate of change in total snowfall from 1930 to 2007 at 419 weather stations in the contiguous 48 states. Blue circles represent increased snowfall; red circles represent a decrease. Data source: Kunkel et al., 2009[15]

Change in Snow-to-Precipitation Ratio in the Contiguous 48 States, 1949–2016

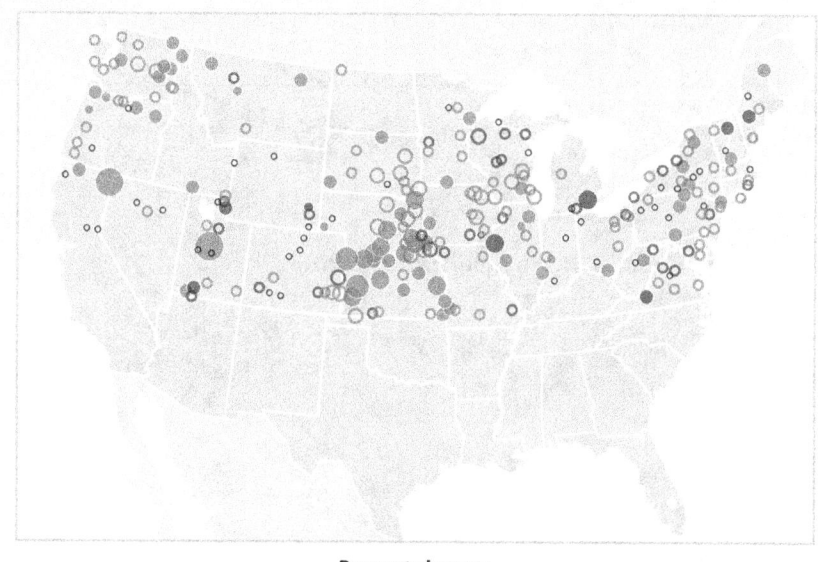

Percent change:

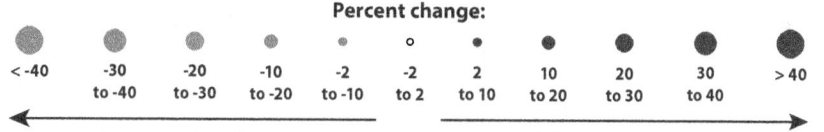

< -40	-30 to -40	-20 to -30	-10 to -20	-2 to -10	-2 to 2	2 to 10	10 to 20	20 to 30	30 to 40	> 40

Lower percentage of snow   Higher percentage of snow

Filled circles represent statistically significant trends.
Open circles represent trends that are not statistically significant.

This figure shows the percentage change in winter snow-to-precipitation ratio from 1949 to 2016 at 246 weather stations in the contiguous 48 states. This ratio measures what percentage of total winter precipitation falls in the form of snow. A decrease (red circle) indicates that more precipitation is falling in the form of rain instead of snow. Solid-color circles represent stations where the trend was statistically significant. Data source: NOAA, 2016[16]

- In addition to changing the overall rate of precipitation, climate change can lead to changes in the type of precipitation. One reason for the decline in total snowfall is because more winter precipitation is falling in the form of rain instead of snow. Nearly 80 percent of the stations across the contiguous 48 states have experienced a decrease in the proportion of precipitation falling as snow.

- Snowfall trends vary by region. The Pacific Northwest has seen a decline in both total snowfall and the proportion of precipitation falling as snow. Parts of the Midwest have also experienced a decrease, particularly in terms of the snow-to-precipitation ratio. A few regions have seen modest increases, including some areas near the Great Lakes that now receive more snow than in the past.

Snow & Ice

ABOUT THE INDICATOR

The graph on p. 48 shows changes in total snowfall, which is determined by the height of snow that accumulates each day. This analysis was adapted from a study by Kunkel et al. (2009).[17] The graph above shows trends in the proportion of total precipitation that falls in the form of snow during each winter season. This is called the "snow-to-precipitation" ratio, and it is based on comparing the amount of snowfall with the total amount of precipitation (snow plus rain) in each year. Both graphs are based on daily records from hundreds of weather stations. These data have been collected and maintained by the National Oceanic and Atmospheric Administration. Stations were selected for this indicator because they had high-quality data for the entire time period of interest.

Snow Cover

This indicator measures the amount of land in North America that is covered by snow.

now cover refers to the amount of land covered by snow at any given time, which is influenced by the amount of precipitation that falls as snow. As temperature and precipitation patterns change, so can the overall area covered by snow. Snow cover is not just something that is affected by climate change, however; it also exerts an influence on climate. More snow means more energy reflects back to space, resulting in cooling, while less snow cover means more energy is absorbed at the Earth's surface, resulting in warming. Some plants and animals may depend on snow to insulate them from sub-freezing winter temperatures, and humans and ecosystems also rely on snowmelt to provide soil moisture and replenish streams and groundwater.

WHAT'S HAPPENING

- When averaged over the entire year, snow covered an average of 3.24 million square miles of North America during the period from 1972 to 2015.

- The extent of snow cover has varied from year to year. The average area covered by snow has ranged from 3.0 million to 3.6 million square miles, with the minimum value occurring in 1998 and the maximum in 1978.

- Between 1972 and 2015, the average extent of North American snow cover decreased at a rate of about 3,300 square miles per year. The average area covered by snow during the most recent decade (2006–2015) was 3.21 million square miles, which is about 4 percent smaller than the average extent during the first 10 years of measurement (1972–1981)—a difference of 122,000 square miles, or approximately an area the size of New Mexico.

Snow-Covered Area in North America, 1972–2015

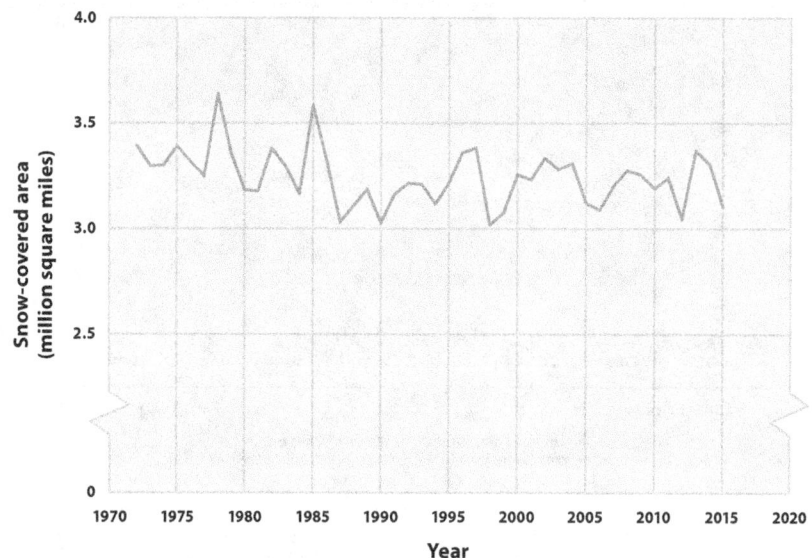

This graph shows the average area covered by snow in a given calendar year, based on an analysis of weekly maps. The area is measured in square miles. These data cover all of North America (not including Greenland). Data source: Rutgers University Global Snow Lab, 2016[18]

Snow Cover Season in the United States, 1972–2013

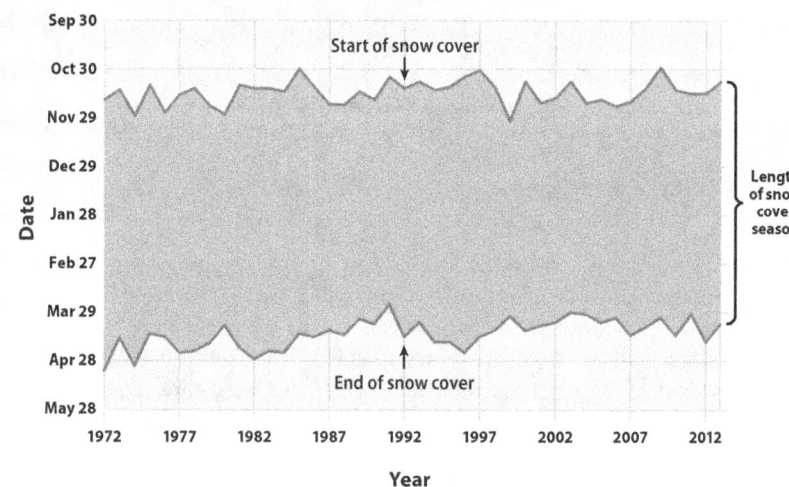

This figure shows the timing of each year's snow cover season in the contiguous 48 states and Alaska, based on an average of all parts of the country that receive snow every year. The shaded band spans from the first date of snow cover until the last date of snow cover. Data source: NOAA, 2015[19]

• Since 1972, the U.S. snow cover season has become shorter by nearly two weeks, on average. By far the largest change has taken place in the spring, with the last day of snow shifting earlier by 19 days since 1972. In contrast, the first date of snow cover in the fall has remained relatively unchanged.

ABOUT THE INDICATOR

This indicator tracks the area covered by snow since 1972, based on maps generated by analyzing satellite images collected by the National Oceanic and Atmospheric Administration. The first graph was created by analyzing each weekly map to determine the extent of snow cover, then averaging the weekly observations together to get a value for each year. This part of the analysis covers all of North America, not including Greenland. The second graph focuses on the contiguous 48 states plus Alaska. It shows the average date when snow first starts to cover the ground in the fall, the average last date of snow cover in the spring, and the length of time between them. These snow cover season dates have been averaged over all parts of the country that regularly receive snow. The online version of this indicator also shows changes in snow-covered area by season over this same time period.

Snow & Ice

Snowpack

This indicator measures trends in mountain snowpack in the western United States.

Temperature and precipitation are key factors affecting snowpack, which is the amount or thickness of snow that accumulates on the ground. Mountain snowpack plays a key role in the water cycle in western North America, storing water in the winter when snow falls and then releasing it as runoff in spring and summer when the snow melts. Millions of people in the West depend on the melting of mountain snowpack for power, irrigation, and drinking water. Changes in mountain snowpack can affect agriculture, winter recreation, and tourism in some areas, as well as plants and wildlife. In a warming climate, more precipitation is expected to fall as rain rather than snow in most areas—reducing the extent and depth of snowpack. Higher temperatures in the spring can cause snow to melt earlier.

ABOUT THE INDICATOR

This indicator examines more than a half-century of snowpack measurements from the United States Department of Agriculture's Natural Resources Conservation Service and the California Department of Water Resources. Snowpack is commonly measured in snow water equivalents, which can be thought of as the depth of water that would result if the entire snowpack were to melt. Snowpack data have been collected over the years using a combination of manual measurements and automated instruments. This indicator shows long-term rates of change for April 1, the most frequent observation date, because it could reflect changes in snowfall and it is extensively used for spring streamflow forecasting.

WHAT'S HAPPENING

- From 1955 to 2016, April snowpack declined at more than 90 percent of the sites measured. The average change across all sites amounts to about a 23-percent decline.

- Large and consistent decreases have been observed throughout the western United States. Decreases have been especially prominent in Washington, Oregon, and the northern Rockies.

- While some stations have seen increases in snowpack, all 11 states included in this indicator have experienced a decrease in snowpack on average over the time period. In the Northwest (Idaho, Oregon, Washington), all but three stations saw decreases in snowpack over the period of record.

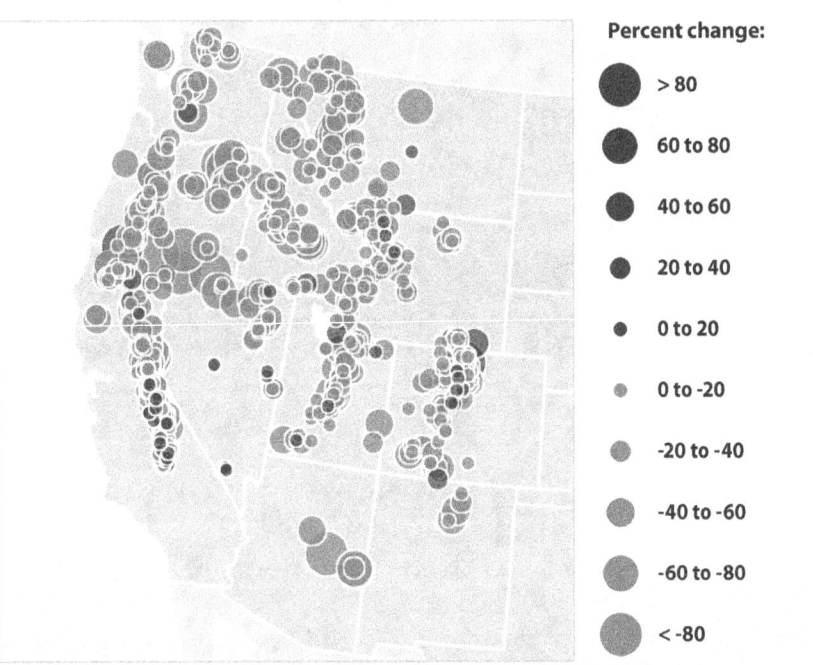

Trends in April Snowpack in the Western United States, 1955–2016

Percent change:
- \> 80
- 60 to 80
- 40 to 60
- 20 to 40
- 0 to 20
- 0 to -20
- -20 to -40
- -40 to -60
- -60 to -80
- < -80

This map shows trends in April snowpack in the western United States, measured in terms of snow water equivalent. Blue circles represent increased snowpack; red circles represent a decrease. Data source: Mote and Sharp, 2016[20]

Understanding the Connections Between
CLIMATE CHANGE
and HUMAN HEALTH

C limate change poses many threats to the health and well-being of Americans, from increasing the risk of extreme heat events and heavy storms to increasing the risk of asthma attacks and changing the spread of certain diseases carried by ticks and mosquitoes. Some of these health impacts are already happening in the United States.

Scientists' understanding of the multiple ways that climate change increases risks to human health has advanced significantly in recent years. This section highlights key concepts from one of the latest climate and health assessments (see the box below) to help illustrate how climate change can affect human health in the United States. In addition, this section demonstrates how EPA's climate change indicators advance the dialogue in connecting climate change and human health.

ACCESS USGCRP'S CLIMATE AND HEALTH ASSESSMENT ONLINE

The U.S. Global Change Research Program (USGCRP) report, *The Impacts of Climate Change on Human Health in the United States: A Scientific Assessment*, was published in April 2016.
This comprehensive report captures the state of scientific knowledge about observed and projected impacts of climate change on human health in the United States. The report is available online at: https://health2016.globalchange.gov.

How Does Climate Change Affect Human Health?

Climate change can exacerbate existing health threats or create new public health challenges through a variety of pathways. Figure 1 summarizes these connections by linking *climate impacts* to changes in *exposure*, which can then lead to negative effects on health *(health outcomes)*. This figure also shows how other factors—such as where people live and their age, health, income, or ability to access health care resources—can positively or negatively influence people's vulnerability to human health effects. For example, a family's income, the quality of their housing, or their community's emergency management plan can all affect that family's exposure to extreme heat, the degree to which their health is affected by this threat, and their ability to adapt to impacts of extreme heat (for more examples, see Figure 4).

Figure 1. Climate Change and Health Pathway

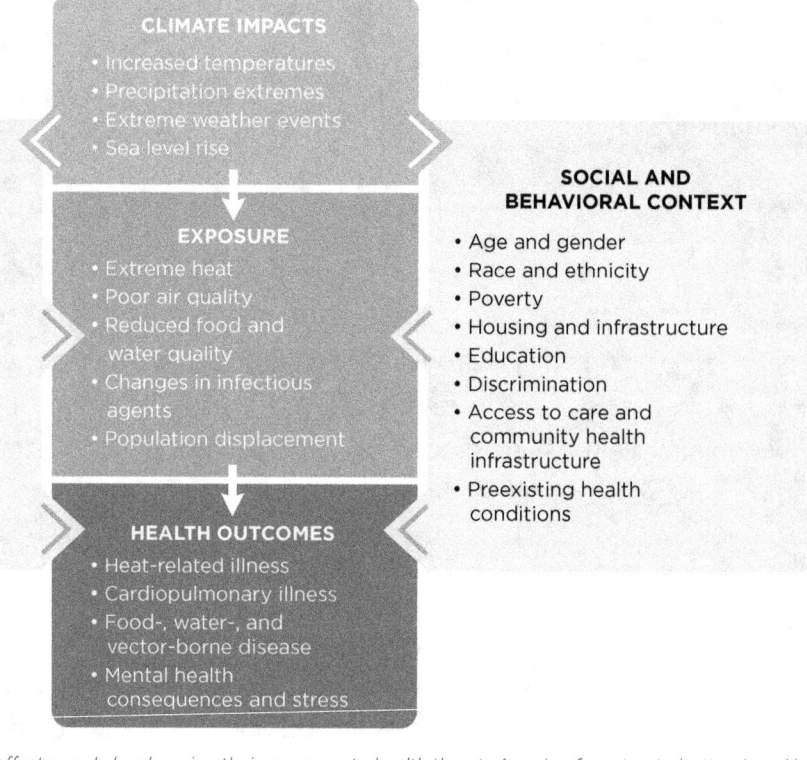

Figure 1 shows how climate change can affect people by changing their exposure to health threats (moving from top to bottom) and by influencing the environmental, institutional, social, and behavioral factors that affect a person's or community's health (moving through the boxes on the sides).

What Can Indicators Tell Us About Climate Change and Human Health?

As shown in Figure 1, the impacts of climate change on health are complex, often indirect, and dependent on multiple societal and environmental factors. Tracking changes in climate impacts and exposures improves understanding of changes in health risk, however, even if the actual health outcome is difficult to quantify. For example, the flooding pathway in Figure 2 shows how indicators of certain climate impacts like Sea Level Rise, Heavy Precipitation, and Coastal Flooding could be used by state and local health officials to better understand changes in human exposure to contaminated waters (a health risk). By recognizing changing risks, these officials can better understand how climate change affects the number of people who get sick with gastrointestinal illnesses (a health outcome). Thus, even where health data or long-term records are unavailable or where the links between climate and health outcomes are complex, indicators play an important role in understanding climate-related health impacts.

Figure 2. Connecting Climate Change Indicators to Health Pathways

The following three examples show how climate impacts can affect health. The numbered circles identify where climate change indica-tors provide key information on changes occurring at different points along the pathways. Other factors can play a role in determining a person's vulnerability to climate-related health outcomes; see Figure 1 and Figure 4.

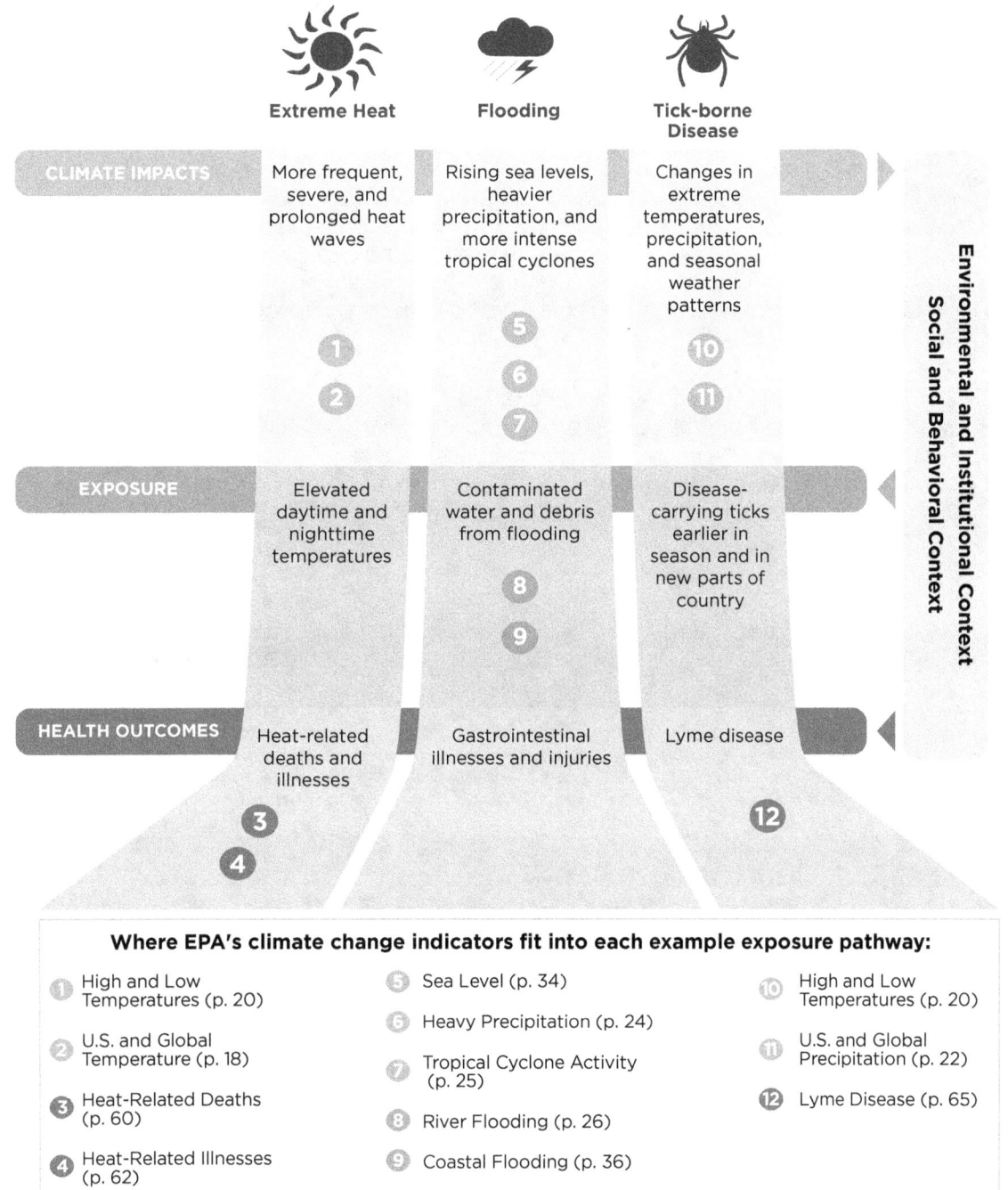

Where EPA's climate change indicators fit into each example exposure pathway:

1. High and Low Temperatures (p. 20)
2. U.S. and Global Temperature (p. 18)
3. Heat-Related Deaths (p. 60)
4. Heat-Related Illnesses (p. 62)

5. Sea Level (p. 34)
6. Heavy Precipitation (p. 24)
7. Tropical Cyclone Activity (p. 25)
8. River Flooding (p. 26)
9. Coastal Flooding (p. 36)

10. High and Low Temperatures (p. 20)
11. U.S. and Global Precipitation (p. 22)
12. Lyme Disease (p. 65)

Who's at Risk?

Every American faces a risk of health impacts associated with climate change. Some people, however, face higher risks than others because of differences in the hazards to which they are exposed, their sensitivity to these hazards, and their ability to adapt (see Figure 3). Thus, it is important to be able to identify "populations of concern," which include groups representing people of all ages, living in

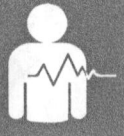

MAKE THE HEALTH CONNECTION
Look for this symbol throughout the report to identify climate change indicators with important health connections.

different places, who interact with their environment in different ways. Figure 4 shows some examples of how certain populations are more vulnerable to health impacts because of differences in their exposure, sensitivity, or ability to adapt to climate-related stresses.

It is important to remember that the different health impacts identified here do not occur in isolation; people can face multiple threats at the same time, at different stages in their lives, or accumulating over the course of their lives. Risks may increase as people are exposed to multiple health threats. For example, extremely hot days can lead to heat-related illness as well as poor air quality, by increasing the chemical reactions that produce smog. In addition, many of the factors that influence whether a person is exposed to health threats or whether they become ill, such as an individual's personal habits, living conditions, and access to medical care (see Figure 1), can also change over time.

The effects of climate change also affect people's mental health. In particular, climate- or weather-related disasters can increase the risk of adverse mental health consequences, especially if they result in damage to homes and livelihoods or loss of loved ones. The mental health impacts of these events can range from minimal stress and distress symptoms to clinical disorders, such as anxiety, depression, and post-traumatic stress.

Figure 3. Determinants of Vulnerabilty

EXPOSURE
Exposure is contact between a person and one or more biological, psychosocial, chemical, or physical stressors, including stressors affected by climate change.

SENSITIVITY
Sensitivity is the degree to which people or communities are affected, either adversely or beneficially, by their exposure to climate variability or change.

ABILITY TO ADAPT
Adaptive capacity is the ability of communities, institutions, or people to adjust to potential hazards such as climate change, to take advantage of opportunities, or to respond to consequences.

VULNERABILITY
of Human Health to Climate Change

HEALTH OUTCOMES
Injury, acute and chronic illness (including mental health and stress-related illness), developmental issues, and death.

Figure 4. Examples of Climate Change Vulnerabilty

EXPOSURE

Low-income populations may be exposed to climate change threats because of socioeconomic factors. For example, people who cannot afford air conditioning are more likely to suffer from unsafe indoor air temperatures.

Occupational groups such as first responders and construction workers face more frequent or longer exposure to climate change threats. For example, extreme heat and disease-carrying insects and ticks particularly affect outdoor workers.

People in certain locations may be exposed to climate change threats, such as droughts, floods, or severe storms, that are specific to where they live. For example, people living by the coast are at increased risk from hurricanes, sea level rise, and storm surge.

SENSITIVITY

Pregnant women are sensitive to health risks from extreme weather such as hurricanes and floods. These events can affect their mental health and the health of their unborn babies by contributing to low birthweight or preterm birth.

People with pre-existing medical conditions, such as asthma, are particularly sensitive to climate change impacts on air quality. People who have diabetes or who take medications that make it difficult to regulate body temperature are sensitive to extreme heat.

Children are more sensitive to respiratory hazards than adults because of their lower body weight, higher levels of physical activity, and still-developing lungs. Longer pollen seasons may lead to more asthma episodes.

ABILITY TO ADAPT

Older adults may have limited ability to cope with extreme weather if, for example, they have difficulty accessing cooling centers or other support services during a heat wave. Heat-related deaths are most commonly reported among adults aged 65 and over.

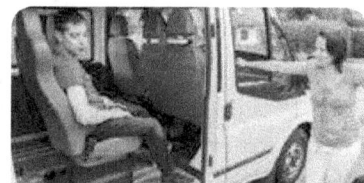

People with disabilities face challenges preparing for and responding to extreme weather events. For example, emergency or evacuation instructions are often not accessible to people with learning, hearing, or visual disabilities.

Indigenous people who rely on subsistence food have limited options to adapt to climate change threats to traditional food sources. Rising temperatures and changes in the growing season affect the safety, availability, and nutritional value of some traditional foods and medicinal plants.

Additional Resources

Climate change threatens human health, including mental health, as well as access to clean air, safe drinking water, nutritious food, and shelter. Understanding the threats that climate change poses to human health can help people and communities work together to lower risks and be prepared.

The following EPA resources on how climate change affects your health can be found at: www.epa.gov/climatechange/impacts:

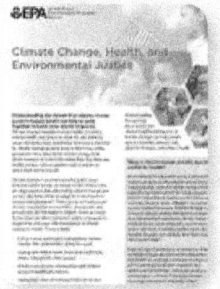

Eight fact sheets on populations shown in Figure 4 that are particularly vulnerable to the health impacts of climate change.

A clickable map with examples of state impacts and resources to help individuals and communities prepare and respond to climate threats.

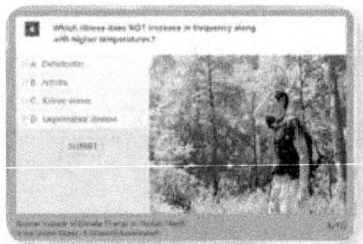

An online 10-question quiz to challenge you and your friends on climate and health knowledge.

A fact sheet highlighting health impacts of climate change at different stages of life, from infancy through adolescence to adulthood.

Health and Society

Changes in the Earth's climate can affect public health, agriculture, water supplies, energy production and use, land use and development, and recreation. The nature and extent of these effects, and whether they will be harmful or beneficial, will vary regionally and over time. This chapter looks at some of the ways that climate change is affecting human health and society, including changes in Lyme disease, West Nile virus, ragweed pollen season, heat-related deaths and hospitalizations, heating and cooling needs, and the agricultural growing season across the United States.

Climate indicators provide key information on changes to environmental exposures and health outcomes (see Understanding the Connections Between Climate Change and Human Health on p. 53). Because impacts on human health are complex, often indirect, and dependent on multiple societal and environmental factors (including how people choose to respond to these impacts), the development of appropriate health-related climate indicators is challenging and still emerging. Even where health data or long-term records are unavailable or where the links between climate and health outcomes are complex, indicators play an important role in understanding climate-related health impacts.

WHY DOES IT MATTER?

Changes in climate affect the average weather conditions to which we are accustomed. These changes may result in multiple threats to human health and welfare. Warmer average temperatures will continue to lead to hotter days and more frequent and longer heat waves, which could increase the number of heat-related illnesses and deaths. Increases in the frequency or severity of extreme weather events, such as storms, increase the risk of dangerous flooding, high winds, and other direct threats to people and property. Warmer temperatures also reduce air quality by increasing the chemical reactions that produce smog, and, along with changes in precipitation patterns and extreme events, could enhance the spread of some diseases.

In addition, climate change could require adaptation on larger and faster scales than in the past, presenting challenges to human well-being and the economy. The more extensively and more rapidly the climate changes, the larger the potential effects on society. The extent to which climate change affects different regions and sectors of society depends not only on the sensitivity of those systems to climate change, but also on their ability to adapt to or cope with climate change. Populations of particular concern include the poor, children, the elderly, those already in poor health, the disabled, and indigenous populations.

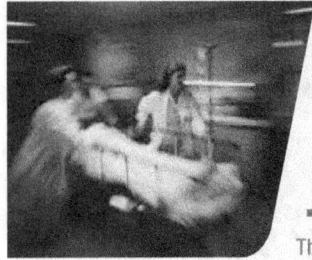

Heat-Related Deaths

This indicator presents data on deaths classified as "heat-related" in the United States.

Unusually hot summer temperatures have become more common across the contiguous 48 states in recent decades[1] (see the High and Low Temperatures indicator on p. 20). When people are exposed to extreme heat, they can suffer from potentially deadly illnesses, such as heat exhaustion and heat stroke. Hot temperatures can also contribute to deaths from heart attacks, strokes, and other forms of cardiovascular disease. Heat is the leading weather-related killer in the United States, even though most heat-related deaths are preventable through outreach and intervention.

Extreme heat events (heat waves) are expected to become longer, more frequent, and more intense in the future.[2] As a result, the risk of heat-related deaths and illness is also expected to increase.[3] Reductions in cold-related deaths are projected to be smaller than increases in heat-related deaths in most regions.[4] Death rates can also change, however, as people acclimate to higher temperatures and as communities strengthen their heat response plans and take other steps to continue to adapt.

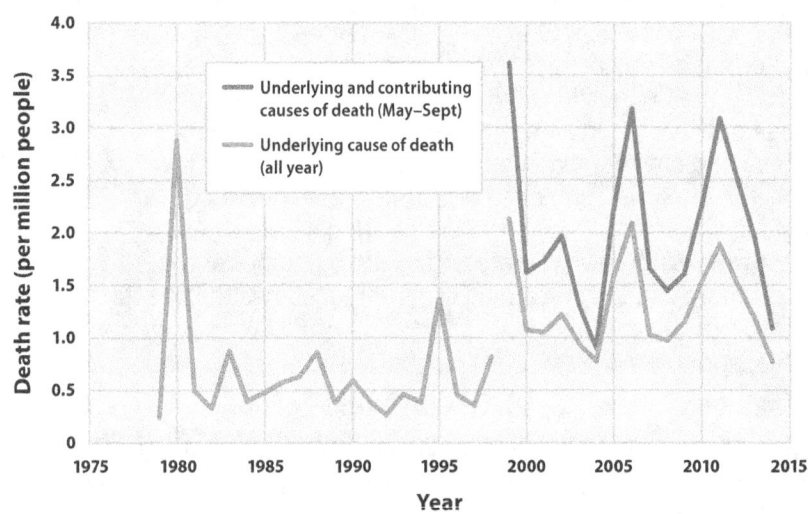

Deaths Classified as "Heat-Related" in the United States, 1979–2014

Legend:
— Underlying and contributing causes of death (May–Sept)
— Underlying cause of death (all year)

This figure shows the annual rates for deaths classified as "heat-related" by medical professionals in the 50 states and the District of Columbia. The orange line shows deaths for which heat was listed as the main (underlying) cause. The blue line shows deaths for which heat was listed as either the underlying or contributing cause of death during the months from May to September, based on a broader set of data that became available in 1999.
Data source: CDC, 2016[5,6]*

* Between 1998 and 1999, the World Health Organization revised the international codes used to classify causes of death. As a result, data from earlier than 1999 cannot easily be compared with data from 1999 and later.

WHAT'S HAPPENING

- Between 1979 and 2014, the death rate as a direct result of exposure to heat (underlying cause of death) generally hovered around 0.5 to 1 deaths per million people, with spikes in certain years. Overall, a total of more than 9,000 Americans have died from heat-related causes since 1979, according to death certificates.

- For years in which the two records overlap (1999–2014), accounting for those additional deaths in which heat was listed as a *contributing* factor results in a higher death rate—nearly double for some years—compared with the estimate that only includes deaths where heat was listed as the *underlying* cause.

- The indicator shows a peak in heat-related deaths in 2006, a year that was associated with widespread heat waves and was one of the hottest years on record in the contiguous 48 states (see the U.S. and Global Temperature indicator on p. 18).

HEALTH CONNECTION

Older adults, particularly those with preexisting health conditions, can be especially vulnerable to extreme heat. Those taking medications that make it difficult to regulate body temperature, who live alone, or who have limited mobility are at higher risk for heat-related illness and death.[7]

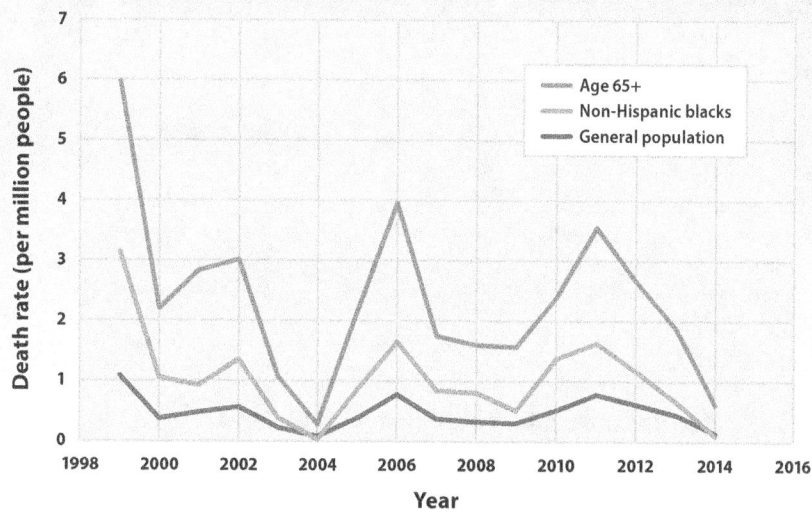

Summer Deaths Due to Heat and Cardiovascular Disease in the United States, 1999–2014

Legend:
- Age 65+
- Non-Hispanic blacks
- General population

This figure shows rates for deaths that medical professionals have classified as being caused by a combination of cardiovascular disease (diseases of the circulatory system) and heat exposure. This graph presents summer (May to September) death rates from 1999 to 2014 for three population groups in the 50 states and the District of Columbia. The purple line shows rates for the entire population, the green line shows rates for non-Hispanic black people, and the pink line shows rates for people aged 65 and older. Data source: CDC, 2016[8]

ABOUT THE INDICATOR

This indicator shows the annual rate for deaths classified by medical professionals as "heat-related" each year in the United States, based on death certificate records compiled by the U.S. Centers for Disease Control and Prevention's National Vital Statistics System. It includes deaths for which excessive natural heat was listed on the death certificate as the main cause of death (also known as the underlying cause), which can be tracked back to 1979. It also examines deaths for which heat was listed as either the underlying cause or a contributing factor, based on a broader set of data that at present can only be evaluated back to 1999. For example, in a case where cardiovascular disease was determined to be the underlying cause of death, heat could be listed as a contributing factor because it can make the individual more susceptible to the effects of this disease. The second graph takes a closer look at heat-related cardiovascular disease deaths, which include deaths due to heart attacks, strokes, and other diseases related to the circulatory system. It shows death rates for the overall population as well as two groups with a higher risk: people aged 65 and older and non-Hispanic blacks.

The numbers shown here do not capture the full extent of heat-related deaths. Many deaths associated with extreme heat are not identified as such by the medical examiner and might not be correctly coded on the death certificate, particularly if they do not occur during an identified or publicized heat event. For example, of the estimated 700 excess deaths during the 1995 heat wave in Chicago, only 465 were recorded and attributed to the extreme heat event (see the online version of this indicator). This type of undercounting is not limited to large heat events. Furthermore, deaths can occur from exposure to heat (either as an underlying cause or as a contributing factor) that is not classified as extreme and therefore is often not recorded as such. Some statistical approaches estimate that more than 1,300 deaths per year in the United States are due to extreme heat, compared with about 600 deaths per year in the "underlying and contributing causes" data set shown in Figure 1.[9]

Health & Society

Heat-Related Illnesses

This indicator tracks how often people are hospitalized because of exposure to heat.

Heat-related illnesses can occur when a person is exposed to high temperatures, such that their body cannot cool itself sufficiently through sweating. Symptoms range from mild swelling, rashes, or cramps to potentially deadly heat exhaustion and heat stroke. Unusually hot summer temperatures have become more common across the contiguous 48 states in recent decades.[10]

Extreme heat events (heat waves) are expected to become longer, more frequent, and more intense in the future.[11] As a result, the risk of heat-related illness is expected to increase.[12] Hospitalization rates can also change, however, as people acclimate to higher temperatures and as communities strengthen their heat response plans and take other steps to continue to adapt.

WHAT'S HAPPENING

- From 2001 to 2010, the 20 states covered in this figure recorded a total of about 28,000 heat-related hospitalizations.[13] The resulting annual rates ranged from 1.1 cases per 100,000 people in 2004 to 2.5 cases per 100,000 people in 2006, with a 10-year average rate of 1.8 cases per 100,000 people.

- The pattern in the figure shown here largely matches the pattern in heat-related deaths during the same period (see the Heat-Related Deaths indicator on p. 60), including a low value in 2004 and a peak in 2006. Considerable year-to-year variability makes it difficult to determine whether heat-related illnesses have increased or decreased to a meaningful degree since 2001.

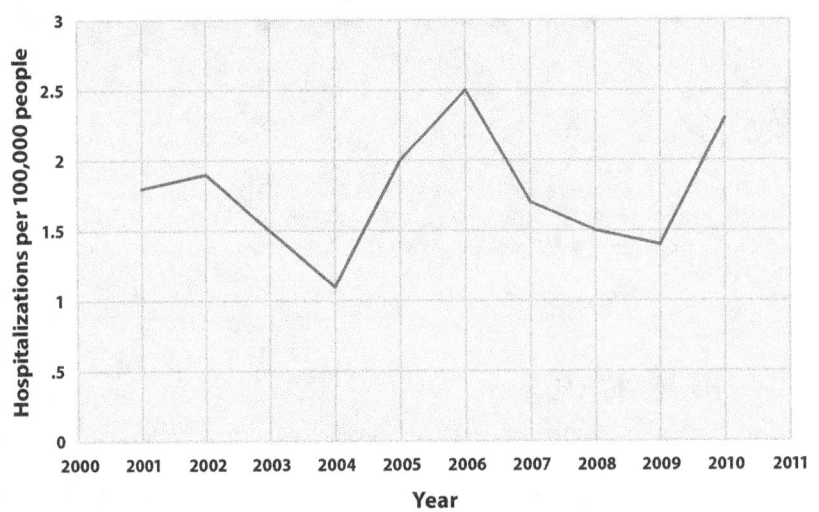

Heat-Related Hospitalizations in 20 States, 2001–2010

This graph shows the annual rate of hospitalizations classified as "heat-related" by medical professionals in 20 states that participate in CDC's hospitalization tracking program, from 2001 to 2010. The rate is based on hospital discharge records for May 1 to September 30 of every year. The rate has been age-adjusted to account for the effects of population change over time—for example, if the proportion of older adults has increased.
Data source: Choudhary and Vaidyanathan, 2014[14]

ABOUT THE INDICATOR

This indicator shows rates for hospital admissions due to "heat-related" illnesses such as heat exhaustion, heat cramps, mild heat edema (swelling in the legs and hands), heat syncope (fainting), and heat stroke. It is based on hospital discharge records, which include a diagnosis determined by a physician or other medical professional. The indicator covers a group of states across a wide range of regions and climate zones that have participated in a national hospital data tracking program since at least 2001. All of these states require hospitals to submit discharge data to a state organization, which then compiles and reports the data to the U.S. Centers for Disease Control and Prevention (CDC). The data for this indicator come from CDC's Environmental Public Health Tracking Program, which includes hospitalization rates per 100,000 people and the total number of heat-related hospitalizations broken out by sex and age group.

Average Rate of Heat-Related Hospitalizations in 23 States, 2001–2010

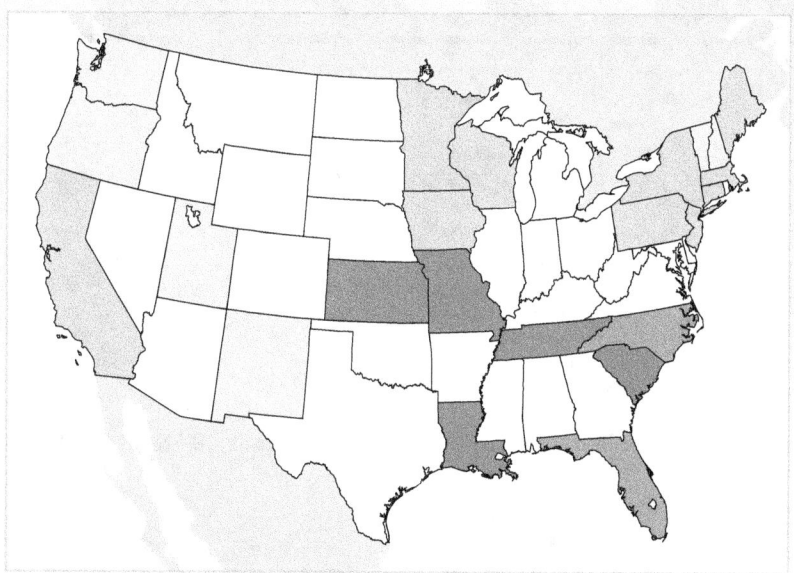

Average annual hospital admissions per 100,000 people:

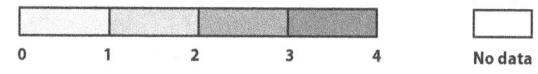

0   1   2   3   4   No data

This map shows the 2001–2010 average rate of hospitalizations classified as "heat-related" by medical professionals in 23 states that participate in CDC's hospitalization tracking program. Rates are based on hospital discharge records for May 1 to September 30 of every year. Rates have been age-adjusted to account for differences in the population distribution over time and between states—for example, if one state has a higher proportion of older adults than another. Data source: CDC, 2016[15]

Heat-Related Hospitalizations in 20 States by Sex and Age, 2001–2010

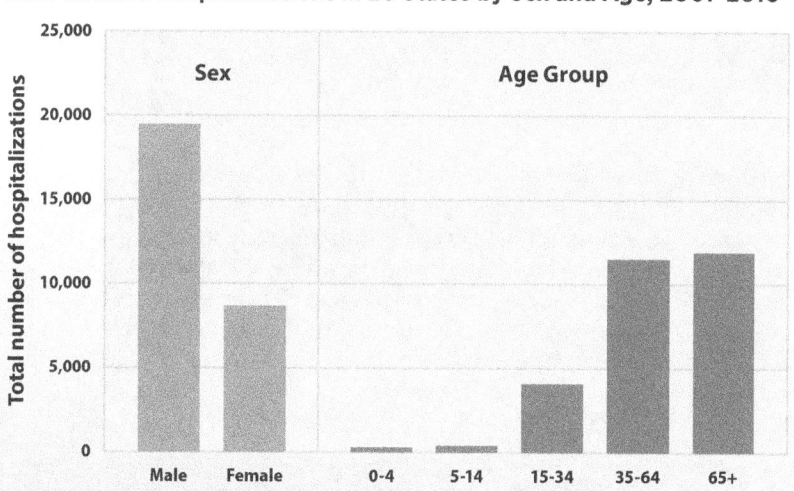

This graph shows the total number of hospitalizations classified as "heat-related" by medical professionals in 20 states that participate in CDC's hospitalization tracking program, from 2001 to 2010. Totals are broken out by sex and by age group. The graph shows 10-year totals based on hospital discharge records for May 1 to September 30 of every year. Data source: Choudhary and Vaidyanathan, 2014[16]

- Heat-related hospitalization rates vary widely among the 23 states studied (see the top figure). Average rates from 2001 to 2010 ranged from fewer than one case per 100,000 people in some states to nearly four cases per 100,000 people in others. The highest rates occurred in Kansas, Louisiana, Missouri, South Carolina, and Tennessee. Relatively high hospitalization rates in the Southeast and Midwest suggest a connection between hotter and more humid summers and increased rates of heat-related illness, compared with other regions.[17]

- People aged 65+ accounted for more heat-related hospitalizations than any other age group from 2001 to 2010, and males were hospitalized for heat-related illnesses more than twice as often as females (see the bottom figure). Men tend to have a higher risk of heat-related illness than women because they are more likely to work in outdoor occupations such as construction.[18]

Health & Society

This indicator does not cover every state, and it could overlook illnesses that were not diagnosed as heat-related, did not result in a hospitalization, or were not fully documented or reported. Nonetheless, this data set represents the best available source of observed data for tracking heat-related hospitalizations across multiple states.

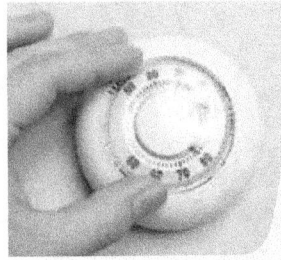

Heating and Cooling Degree Days

This indicator examines changing temperatures from the perspective of heating and cooling needs for buildings.

As climate change contributes to an increase in average temperatures, an increase in unusually hot days, and a decrease in unusually cold days (see the U.S. and Global Temperature and High and Low Temperatures indicators on pp. 18 and 20), the overall demand for heating is expected to decline, and the demand for cooling is expected to increase. One way to measure the influence of temperature change on energy demand is using heating and cooling degree days, which measure the difference between outdoor temperatures and a temperature that people generally find comfortable indoors. These measurements suggest how much energy people might need to use to heat and cool their homes and workplaces, thus providing a sense of how climate change could affect people's daily lives and finances.

WHAT'S HAPPENING

- Heating degree days have declined in the contiguous United States, particularly in recent years, as the climate has warmed. This change suggests that heating needs have decreased overall.

- Overall, cooling degree days have increased over the past 100 years. The increase is most noticeable over the past few decades, suggesting that air-conditioning energy demand has also been increasing recently.

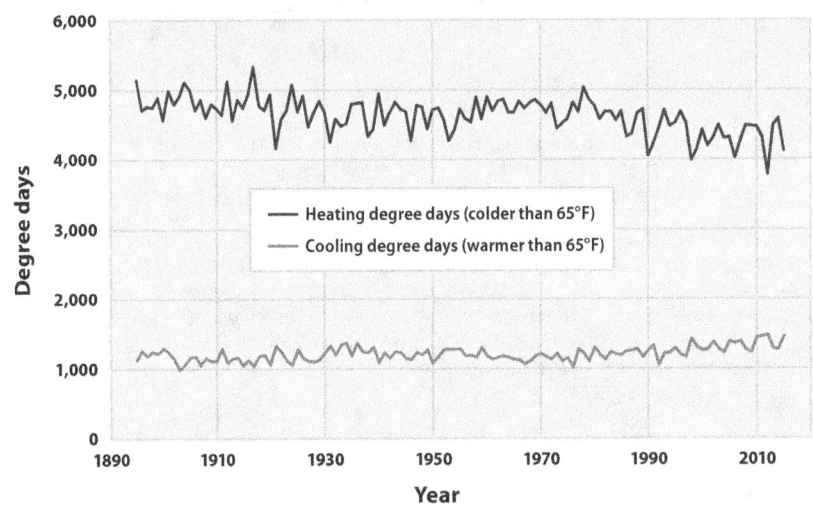

Heating and Cooling Degree Days in the Contiguous 48 States, 1895–2015

This figure shows the average number of heating and cooling degree days per year across the contiguous 48 states. Data source: NOAA, 2016[19]

ABOUT THE INDICATOR

This indicator uses daily temperature data from thousands of weather stations across the contiguous 48 states to calculate heating and cooling degree days. A "degree day" is determined by comparing the daily average outdoor temperature with a defined baseline temperature for indoor comfort (in this case, 65°F). For example, if the average temperature on a particular day is 78°F, then that day counts as 13 cooling degree days, as a building's interior would need to be cooled by 13°F to reach 65°F. Conversely, if the average outdoor temperature is 34°F, then that day counts as 31 heating degree days, as a building's interior would need to be warmed by 31°F to reach 65°F. This does not mean that all people will actually heat or cool buildings to 65°F; it is just a number to allow for consistent comparisons over time and across the country.

The graph above was created by calculating the total number of heating and cooling degree days per year at each weather station, averaging the results from all stations within small regions called climate divisions, then calculating a national average weighted by the population of each climate division. This population-weighting approach produces a national average that more closely reflects the conditions that the average resident would experience. Data and analyses were provided by the National Oceanic and Atmospheric Administration. The online version of this indicator also shows a map with changes in annual heating and cooling degree days by state.

Lyme Disease

This indicator tracks the rate of reported Lyme disease cases across the United States.

Lyme disease is a bacterial illness transmitted through the bite of certain species of ticks (commonly known as deer ticks). It can cause fever, fatigue, joint pain, and skin rash, as well as more serious joint and nervous system complications. Warming temperatures are projected to expand the range of suitable tick habitat,[20] increasing the potential risk of Lyme disease. Also, because deer ticks are mostly active when temperatures are above 45°F, shorter winters could extend the period when ticks are active each year.[21] Climate is not the only factor, however, that could influence the transmission, distribution, and incidence of Lyme disease. Other factors include changes in the populations of host species such as deer and white-footed mice, habitat changes, and the extent to which people take precautions to avoid getting infected.

Reported Cases of Lyme Disease in the United States, 1991–2014

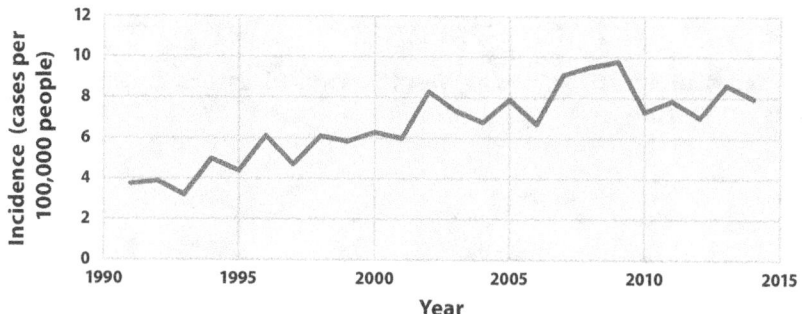

This figure shows the annual incidence of Lyme disease, which is calculated as the number of new cases per 100,000 people. The graph is based on cases that local and state health departments report to the Centers for Disease Control and Prevention's national disease tracking system. Data source: CDC, 2015[22]

- The incidence of Lyme disease in the United States has approximately doubled since 1991, from 3.74 reported cases per 100,000 people to 7.95 reported cases per 100,000 people in 2014.

- Driven by multiple factors, the number and distribution of reported cases of Lyme disease have increased over time.

Reported Lyme Disease Cases in 1996 and 2014

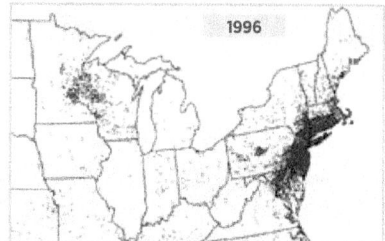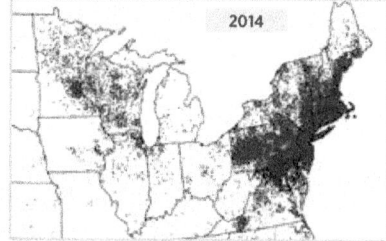

These maps show the distribution of reported cases of Lyme disease in 1996 and 2014. Each dot represents an individual case placed according to the patient's county of residence, which may be different than the county of exposure. The year 1996 was chosen as a reasonable starting point for comparison with recent years. These maps focus on the parts of the United States where Lyme disease is most common. Data source: CDC, 2015[23]

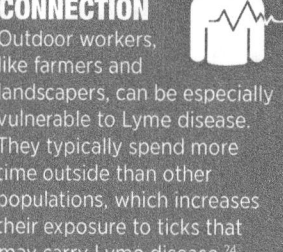

HEALTH CONNECTION

Outdoor workers, like farmers and landscapers, can be especially vulnerable to Lyme disease. They typically spend more time outside than other populations, which increases their exposure to ticks that may carry Lyme disease.[24]

Health & Society

ABOUT THE INDICATOR

This indicator examines the rate of confirmed Lyme disease cases nationwide over time. It is based on data collected by state and local health departments, which track confirmed cases of Lyme disease that are diagnosed by health care providers. These agencies report total cases to the U.S. Centers for Disease Control and Prevention, which compiled the national statistics shown above. Nationwide reporting of Lyme disease began in 1991. The actual number of illnesses is likely greater than what is reported to health officials and shown here, considering that Lyme disease can be difficult to diagnose.[25] The online version of this indicator also shows a map of changes in reported Lyme disease incidence by state.

West Nile Virus

This indicator tracks the rate of reported West Nile virus disease cases across the United States.

West Nile virus is the most common cause of mosquito-borne disease in the United States in most years. Infection with the virus can lead to symptoms such as headaches, body aches, joint pains, vomiting, diarrhea, and rash, as well as more severe damage to the central nervous system in some patients, causing encephalitis, meningitis, and occasionally death.[26] Climate change may raise the risk of human exposure to West Nile virus, which is transmitted between birds and mosquitoes and causes human disease when infected mosquitoes bite people. Studies show that warmer temperatures associated with climate change can speed up mosquito development, biting rates, and the incubation of the disease within a mosquito.[27] Mild winters and drought have also been associated with West Nile virus disease outbreaks.[28,29] Climate change's effects on birds, the main hosts of the virus, may also contribute to changes in long-range virus movement, as the timing of migration and breeding patterns are driven by climate.

WHAT'S HAPPENING

- The incidence of West Nile virus neuroinvasive disease in the United States has varied widely from year to year. No obvious long-term trend can be detected yet through this limited data set.

- The years 2002, 2003, and 2012 had the highest reported incidence rates, around one case per 100,000 people.

Reported Neuroinvasive Cases of West Nile Virus Disease in the United States, 2002–2014

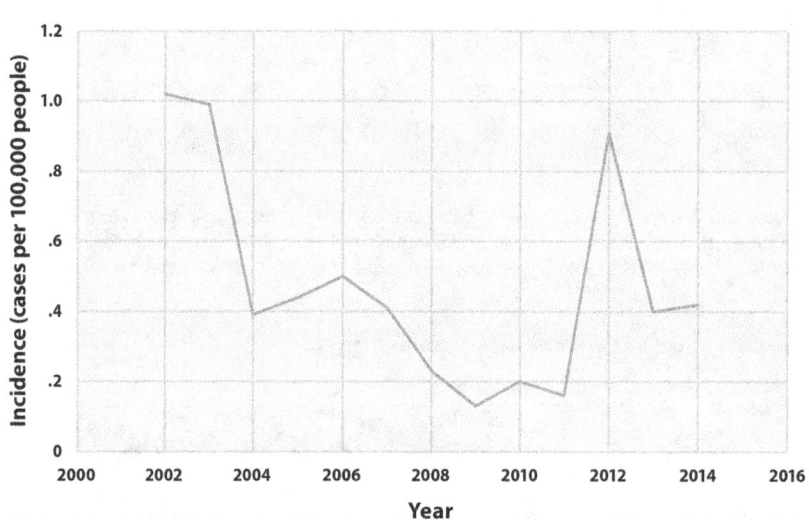

This figure shows the annual incidence of West Nile virus neuroinvasive disease, which is calculated as the number of new cases per 100,000 people. The graph is based on cases that local and state health departments report to the Centers for Disease Control and Prevention's national disease tracking system. Neuroinvasive cases, which occur for less than 1 percent of people infected with West Nile virus, are those that affect the brain or cause neurologic dysfunction. Data source: CDC, 2016[30]

- West Nile virus occurs throughout the contiguous 48 states. Average annual incidence is highest in parts of the Southwest, the Mississippi Delta region, the Great Plains, and the Rocky Mountain region.

Reported Neuroinvasive Cases of West Nile Virus Disease by State, 2002–2014

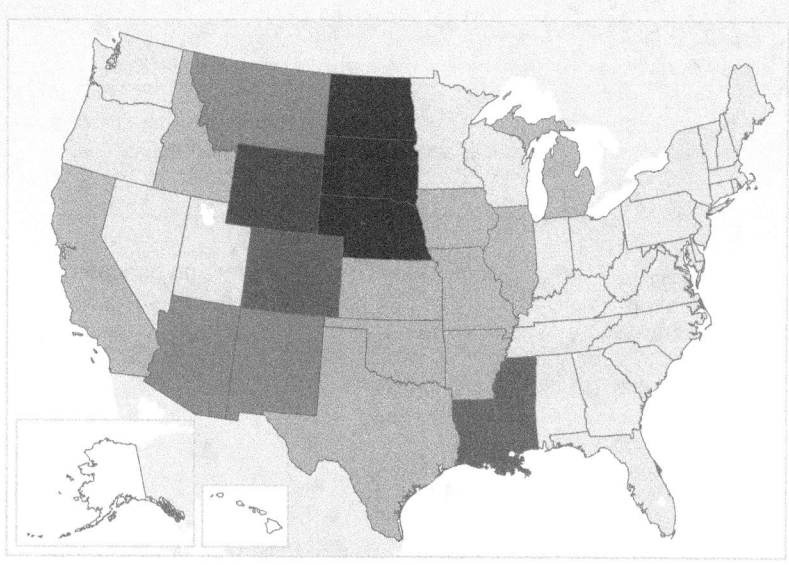

Average annual cases per 100,000 people:

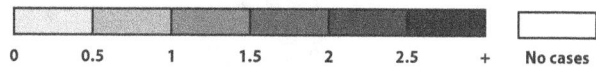

| 0 | 0.5 | 1 | 1.5 | 2 | 2.5 | + | No cases |

This map shows the average annual incidence of West Nile virus neuroinvasive disease in each state, which is calculated as the average number of new cases per 100,000 people per year from 2002 to 2014. The map is based on cases that local and state health departments report to the Centers for Disease Control and Prevention's national disease tracking system. Neuroinvasive cases, which occur for less than 1 percent of people infected with West Nile virus, are those that affect the brain or cause neurologic dysfunction.
Data source: CDC, 2016[31]

HEALTH CONNECTION
People who work or spend large amounts of time outdoors, like farmers, military personnel, or homeless people, can be especially vulnerable to West Nile virus.[32,33] A higher risk of West Nile virus infection is also associated with advanced age and being male.[34]

ABOUT THE INDICATOR

This indicator looks at the incidence of human cases of West Nile virus disease. It focuses on neuroinvasive cases, as the symptoms are noticeable and typically require medical care, which makes detection and reporting more consistent. West Nile became a nationally notifiable disease in 2002, which means health care providers are required to report confirmed cases to their local or state health departments. The U.S. Centers for Disease Control and Prevention compiles these reported data and calculates national and state-level totals and rates.

Length of Growing Season

This indicator measures the length of the growing season in the contiguous 48 states.

The length of the growing season in any given region refers to the number of days when plant growth takes place. The growing season often determines which crops can be grown in an area, as some crops require long growing seasons, while others mature rapidly. Depending on the region and the climate, the growing season is influenced by air temperatures, frost days, rainfall, or daylight hours. Changes in the length of the growing season can have both positive and negative effects on the yield and prices of particular crops. Overall, warming is expected to have negative effects on yields of major crops, but crops in some individual locations may benefit.[35] A longer growing season could also disrupt the function and structure of a region's ecosystems and could, for example, alter the range and types of animal species in the area.

WHAT'S HAPPENING

- The length of the growing season for crops has increased in almost every state. States in the Southwest (e.g., Arizona and California) have seen the most dramatic increase. In contrast, the growing season has actually become shorter in a few southeastern states.

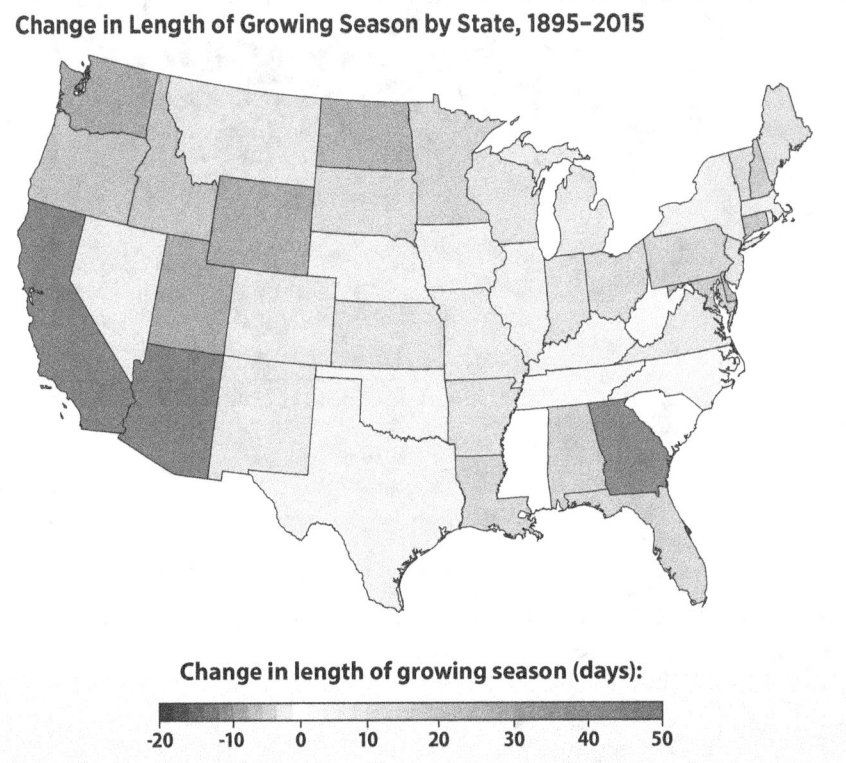

Change in Length of Growing Season by State, 1895–2015

Change in length of growing season (days):

-20 -10 0 10 20 30 40 50

This map shows the total change in length of the growing season from 1895 to 2015 for each of the contiguous 48 states. Data source: Kunkel, 2016[36]

ABOUT THE INDICATOR

For this indicator, the length of the growing season is defined as the period of time between the last frost of spring and the first frost of fall, when the air temperature drops below the freezing point of 32°F. Temperature measurements come from weather stations in the National Oceanic and Atmospheric Administration's Cooperative Observer Program. Growing season length was averaged by state, while the timing of spring and fall frosts were averaged across the nation, then compared with long-term average numbers (1895–2015) to determine how each year differed from the long-term average. The online version of this indicator provides additional maps and graphs that track the length of the growing season nationwide and changes in the timing of the last spring frost and first fall frost by state.

Timing of Last Spring Frost and First Fall Frost in the Contiguous 48 States, 1895–2015

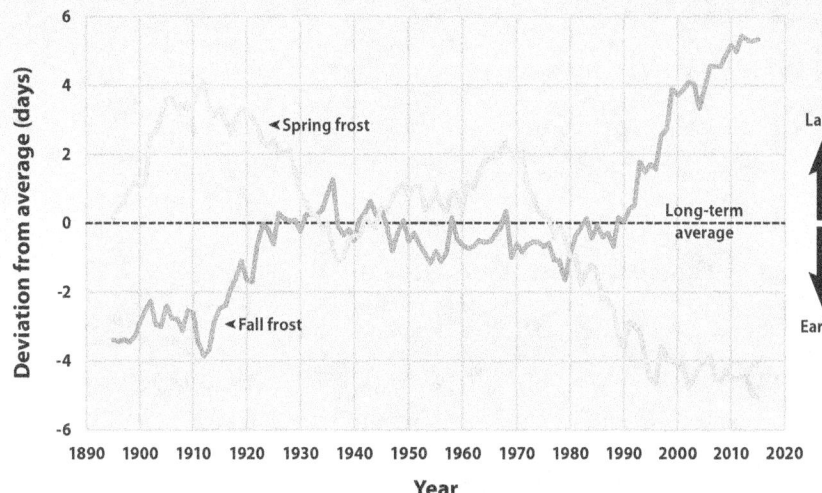

This figure shows the timing of the last spring frost and the first fall frost in the contiguous 48 states compared with a long-term average. Positive values indicate that the frost occurred later in the year, and negative values indicate that the frost occurred earlier in the year. The lines were smoothed using an 11-year moving average. Choosing a different long-term average for comparison would not change the shape of the data over time. Data source: Kunkel, 2016[37]

- In recent years, the final spring frost has been occurring earlier than at any point since 1895, and the first fall frost has been arriving later. Since 1980, the last spring frost has occurred an average of three days earlier than the long-term average, and the first fall frost has occurred about three days later.

Health & Society

Ragweed Pollen Season

This indicator depicts changes in the length of ragweed pollen season in the United States and Canada.

One of the most common environmental allergens is ragweed, which can cause hay fever and trigger asthma attacks. Ragweed pollen season usually peaks in late summer and early fall, but these plants often continue to produce pollen until the first frost. Climate change can affect pollen allergies in several ways. Warmer spring temperatures cause some plants to start producing pollen earlier (see the Leaf and Bloom Dates indicator on p. 82), while warmer fall temperatures extend the growing season for other plants, such as ragweed (see the Length of Growing Season indicator on p. 68). Warmer temperatures and increased carbon dioxide concentrations also enable ragweed and other plants to produce more pollen.[38] This means that many locations could experience longer allergy seasons and higher pollen counts as a result of climate change.

WHAT'S HAPPENING

- Since 1995, ragweed pollen season has grown longer at 10 of the 11 locations studied.

- The increase in ragweed season length generally becomes more pronounced from south to north. Ragweed season increased by 25 days in Winnipeg, Manitoba; 24 days in Saskatoon, Saskatchewan; 21 days in Fargo, North Dakota; and 18 days in Minneapolis, Minnesota. This trend is consistent with many other observations showing that climate is changing more rapidly at higher latitudes.[39]

- The trends shown are strongly related to changes in the length of the frost-free season and the timing of the first fall frost. Northern areas have seen fall frosts happening later than they used to, with the delay in first frost closely matching the increase in pollen season. Meanwhile, some southern stations have experienced only a modest change in frost-free season length since 1995.[40]

Change in Ragweed Pollen Season, 1995–2015

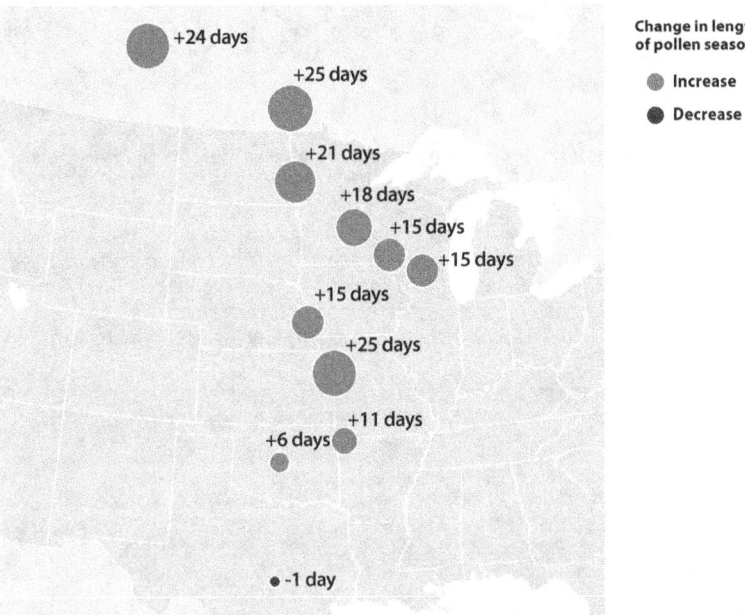

This figure shows how the length of ragweed pollen season changed at 11 locations in the central United States and Canada between 1995 and 2015. Red circles represent a longer pollen season; the blue circle represents a shorter season. Larger circles indicate larger changes. Data source: Ziska et al., 2016[41]

ABOUT THE INDICATOR

This indicator shows changes in the length of the ragweed pollen season in 11 cities that were selected as part of a study that looked at trends in pollen season at sites similar in elevation and across a range of latitudes from south to north. Air samples have been collected and examined at each location since at least the 1990s. Pollen spores are counted and identified using microscopes. Pollen count data have been compiled by the National Allergy Bureau in the United States and Aerobiology Research Laboratories in Canada. Trends were analyzed by a team of researchers that published a more detailed version of this analysis in 2011.[43]

HEALTH CONNECTION

Longer pollen seasons increase people's exposure to pollen and may lead to more asthma episodes and other allergy-related illnesses, especially for children. Children are more sensitive than adults to the effects of pollen and other respiratory hazards because of their level of physical activity and body weight, and because their lungs continue to develop through adolescence.[42]

Ecosystems

Ecosystems provide humans with food, clean water, and a variety of other services that can be affected by climate change. This chapter looks at some of the ways that climate change affects ecosystems, including changes in wildfires, streams and lakes, bird migration patterns, fish and shellfish populations, and plant growth.

WHY DOES IT MATTER?

Changes in the Earth's climate can affect ecosystems by altering the water cycle, habitats, animal behavior—such as nesting and migration patterns—and the timing of natural processes such as flower blooms. Changes that disrupt the functioning of ecosystems may increase the risk of harm or even extinction for some species. While wildfires occur naturally, for example, more frequent and more intense fires can significantly disrupt ecosystems, damage property, put people and communities at risk, and create air pollution problems even far away from the source.

While plants and animals have adapted to environmental change for millions of years, the climate changes being experienced now could require adaptation on larger and faster scales than current species have successfully achieved in the past, potentially increasing the risk of extinction or severe disruption for many species.

Ecosystems

Wildfires

This indicator tracks the extent of wildfires in the United States.

Although wildfires occur naturally and play a long-term role in the health of forests, shrublands, and grassland, climate change threatens to increase the frequency, extent, and severity of fires through numerous factors, such as increased temperatures and drought (see the U.S. and Global Temperature and Drought indicators on pp. 18 and 28). Wildfires have the potential to harm property, livelihoods, and human health. Beyond the human impact, wildfires also affect the Earth's climate. Forests in particular store large amounts of carbon. When they burn, they release carbon dioxide into the atmosphere, which in turn contributes to climate change.

ABOUT THE INDICATOR

The figures here show the total land area burned nationwide and by state. Data for the graph come from the National Interagency Fire Center, which compiles reports from local, state, and federal agencies that are involved in fighting wildfires. The U.S. Forest Service tracked similar data using a different reporting system until 1997. Those data have been added to the graph for comparison. Data for the map come from the Monitoring Trends in Burn Severity project, sponsored by the Wildland Fire Leadership Council. This project uses satellite images taken before and after wildfires to assess the severity of damage. Other parts of this indicator available online track the total number of fires (frequency) and the degree of damage that fires cause to the landscape (severity).

WHAT'S HAPPENING

- The extent of area burned by wildfires each year appears to have increased since the 1980s. According to National Interagency Fire Center data, of the 10 years with the largest acreage burned, nine have occurred since 2000, including the peak year in 2015. This period coincides with many of the warmest years on record nationwide (see the U.S. and Global Temperature indicator on p. 18).

Wildfire Extent in the United States, 1983–2015

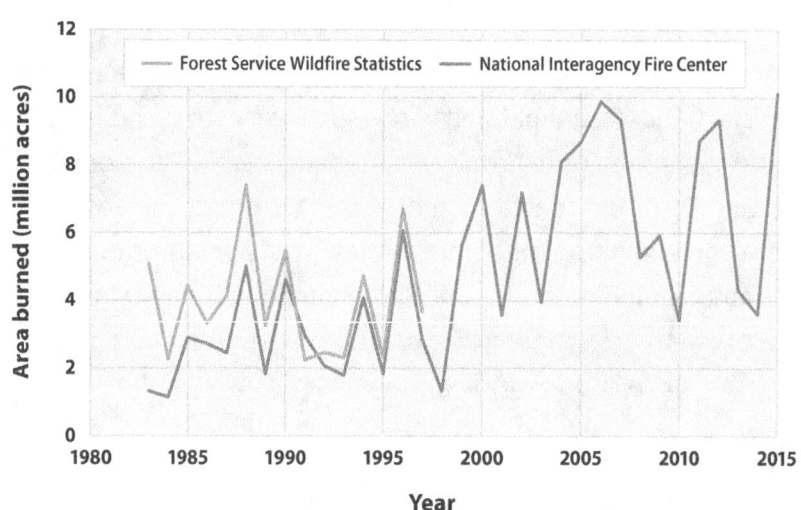

This figure shows annual wildfire-burned area (in millions of acres) from 1983 to 2015. The two lines represent two different reporting systems; though the Forest Service stopped collecting statistics (orange line) in 1997 and is not planning to update them, those statistics are shown here for comparison. Data source: NIFC, 2016;[1] Short, 2015[2]

Average Annual Burned Acreage by State, 1984–2014

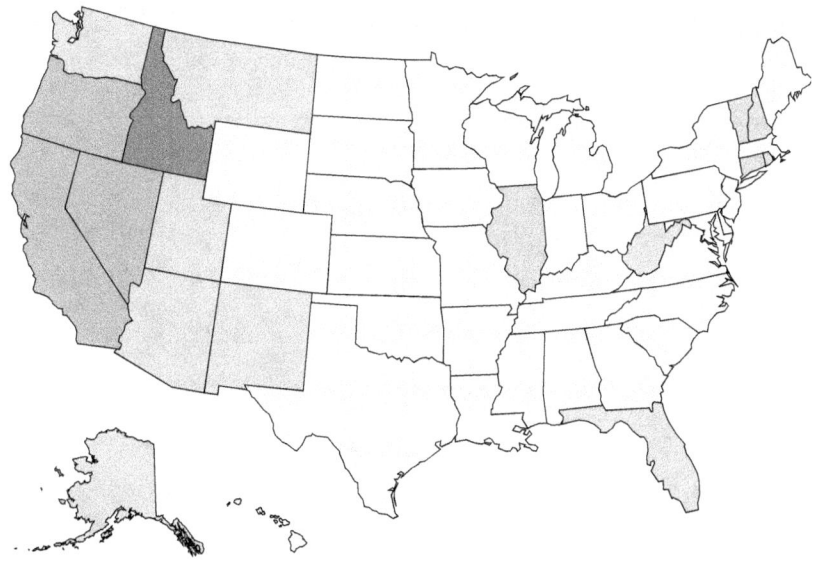

Change in annual burned acreage:

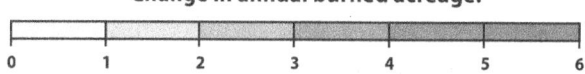

0   1   2   3   4   5   6

States colored light gray did not have any fires that were large enough
to be included in this analysis.

This map shows the average number of acres burned in each state per year as a proportion of that state's total land area. Darker-shaded states have the largest proportion of acreage burned. For reference, there are 640 acres in a square mile; therefore, an average burned area of 6.4 acres per square mile would mean that fires burned 1 percent of a state's total land area. A few states did not have any fires that were large enough to be included in this analysis. Visit this indicator online at: www.epa.gov/climate-indicators for an interactive version of this map. Data source: MTBS, 2016[3]

- Land area burned by wildfires varies by state. Fires burn more land in the western United States than in the East.

HEALTH CONNECTION
Wildfires worsen air quality. Fine particles present in wildfire smoke can drift many miles away from the site of the fire. These air pollutants increase the risk of premature death as well as chronic and acute cardiovascular and respiratory health problems.[4]

Ecosystems

73

Streamflow

This indicator describes trends in the amount of water carried by streams across the United States, as well as the timing of runoff associated with snowmelt.

Streamflow is a measure of the rate at which water is carried by rivers and streams, and it represents a critical resource for people and the environment. Climate change can affect streamflow in several ways. For example, changes in the amount of spring snowpack (see the Snowpack indicator on p. 52) and air temperatures that influence melting can alter the size and timing of high spring streamflows. Changes in precipitation and drought patterns could increase or reduce streamflow in certain areas. Changes in streamflow can directly influence the supply of drinking water and the amount of water available for irrigating crops, generating electricity, and other needs. In addition, many plants and animals depend on streamflow for habitat and survival.

Seven-Day Low Streamflows in the United States, 1940–2014

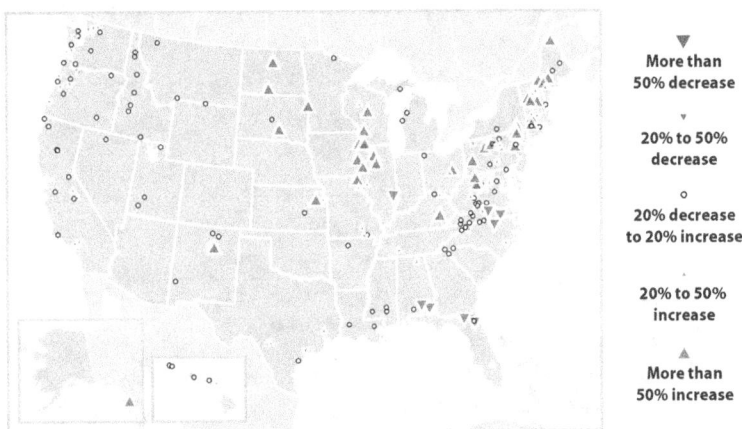

▼ More than 50% decrease

▼ 20% to 50% decrease

○ 20% decrease to 20% increase

▲ 20% to 50% increase

▲ More than 50% increase

This map shows percentage changes in the minimum annual streamflow for rivers and streams across the country, based on the long-term rate of change from 1940 to 2014. Minimum streamflow is based on the consecutive seven-day period with the lowest average flow during a given year. Data source: USGS, 2016[5]

Three-Day High Streamflows in the United States, 1940–2014

▼ More than 50% decrease

▼ 20% to 50% decrease

○ 20% decrease to 20% increase

▲ 20% to 50% increase

▲ More than 50% increase

This map shows percentage changes in the maximum annual streamflow for rivers and streams across the country, based on the long-term rate of change from 1940 to 2014. Maximum streamflow is based on the consecutive three-day period with the highest average flow during a given year. Data source: USGS, 2016[6]

WHAT'S HAPPENING

- During the past 75 years, seven-day low flows have generally increased in the Northeast and Midwest (in other words, on the days of lowest flows, streams in these areas are carrying more water than before). Low flows have generally decreased in parts of the Southeast and the Pacific Northwest (that is, streams are carrying less water than before). Overall, more sites have experienced increases than decreases.

- Three-day high-flow trends vary from region to region across the country. For example, high flows have generally increased or changed little in the Northeast since 1940, whereas high flows have increased in some West Coast streams and decreased in others. Overall, more sites have experienced increases than decreases.

Annual Average Streamflow in the United States, 1940–2014

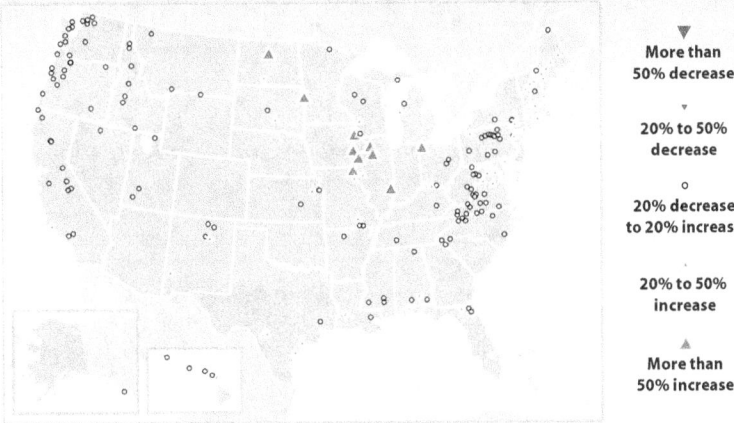

More than
50% decrease

20% to 50%
decrease

20% decrease
to 20% increase

20% to 50%
increase

More than
50% increase

This map shows percentage changes in the annual average streamflow for rivers and streams across the country, based on the long-term rate of change from 1940 to 2014. This map is based on daily streamflow measurements, averaged over the entire year.
Data source: USGS, 2016[7]

Timing of Winter-Spring Runoff in the United States, 1940–2014

More than
10 days earlier

5 to 10
days earlier

2 to 5
days earlier

2 days earlier
to 2 days later

2 to 5
days later

5 to 10
days later

More than
10 days later

This map shows changes in the timing of annual high winter-spring flow carried by rivers and streams from 1940 to 2014. This analysis focuses on parts of the country where streamflow is strongly influenced by snowmelt. Trends are based on the winter-spring center of volume, which is the date when half of the total January 1–July 31 streamflow (in the West) or half of the total January 1–May 31 streamflow (in the East) has passed by each streamflow gauge.
Data source: USGS, 2016[8]

- Annual average streamflow has increased at many sites in the Northeast and Midwest, while other regions have seen few substantial changes. Overall, sites show more increases than decreases.

- In parts of the country with substantial snowmelt, winter-spring runoff is happening more than five days earlier than in the mid-20th century at most gauges. The largest changes occurred in the Pacific Northwest and Northeast.

ABOUT THE INDICATOR

This indicator is based on measurements taken by the U.S. Geological Survey using continuous monitoring devices called stream gauges. The indicator focuses on sites where trends are not substantially influenced by dams, reservoir management, wastewater treatment facilities, or land-use change. The lowest flows each year are commonly calculated by averaging the lowest seven consecutive days of streamflow, while the highest flows each year are commonly calculated by averaging the highest three consecutive days of streamflow. Annual average streamflow is calculated by averaging daily flows through the entire year. The fourth graph examines the timing of winter and spring runoff in areas where at least 30 percent of annual precipitation falls as snow. Scientists look at the total volume of water that passes by a gauge between January 1 and July 31 for the western United States, and January 1 and May 31 for the eastern United States, then determine the date when exactly half of that water has gone by. This date is called the winter-spring center-of-volume date.

Ecosystems

Stream Temperature

This indicator shows changes in stream water temperature across the Chesapeake Bay region.

Rising air temperatures (see the U.S. and Global Temperature indicator on p. 18), along with other factors such as land-use changes, can contribute to higher water temperatures in streams. This warming can affect water quality and aquatic life. Many plants, animals, and other organisms living in streams can flourish only in a specific range of water temperatures. Higher temperatures reduce levels of dissolved oxygen in the water, which can negatively affect the growth and productivity of aquatic life, and can accelerate natural chemical reactions and release excess nutrients into the water.[9] A stream's water temperature can also influence the circulation or mixing patterns in the water it flows into, like bays and estuaries, potentially affecting nutrient levels and salinity. The Chesapeake Bay is the largest estuary in the United States, an important habitat for countless aquatic species, and a driver of the regional economy. Warmer stream water coming into the bay can stress plants and animals and worsen the effects of nutrient pollution that the bay is already facing.[10]

WHAT'S HAPPENING

- From 1960 through 2014, water temperature increased at 79 percent of the stream sites measured in the Chesapeake Bay region. More than half of these increases were statistically significant. Only 5 percent of stations had a significant temperature decrease over the same period.

- Since 1960, the Chesapeake Bay region has experienced an overall increase in stream water temperature. Temperature has risen by an average of 1.2°F across all sites and 2.2°F at the sites where trends were statistically significant.

- Stream temperatures have risen throughout the Chesapeake Bay region. The largest increases have occurred in the southern part of the region.

Changes in Stream Water Temperatures in the Chesapeake Bay Region, 1960–2014

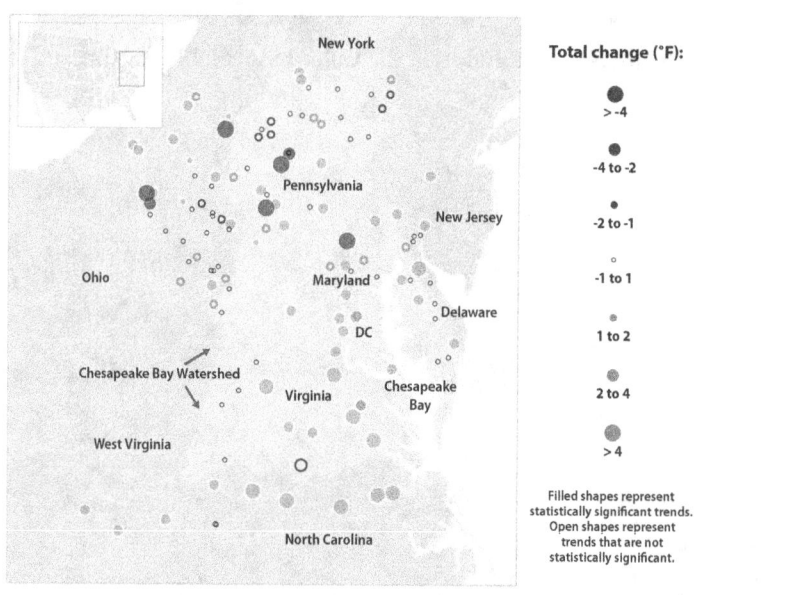

This map shows the change in water temperature at 129 stream gauges across the Chesapeake Bay region from 1960 to 2014. Red circles show locations where temperatures have increased; blue circles show locations where temperatures have decreased. Solid-color circles represent sites where the change was statistically significant. Data source: Jastram and Rice, 2015[11]

ABOUT THE INDICATOR

This indicator is based on an analysis developed by the U.S. Geological Survey. It uses water temperature data from a set of stream gauging stations in the Chesapeake Bay region. Field technicians visit each gauging station an average of eight times a year to measure various stream conditions, including water temperature. The data were analyzed in a way that accounts for variations in timing and makes it possible to compare average temperatures across many years. In addition to climate, changes to a stream's average water temperature can be influenced by other factors such as industrial discharges, changes to local hydrology (such as construction and operation of dams and channels), and changes to land cover in the watershed (including the amount of shade that trees provide to the stream). Nonetheless, this study found that sites without many of these complicating factors warmed just as much as sites with more extensive human influence.[12]

TRIBAL CONNECTION: TRENDS IN STREAM TEMPERATURE IN THE SNAKE RIVER

Climate change has challenged and will continue to challenge some of the traditional ways of life that have sustained indigenous peoples for thousands of years. In the Pacific Northwest, warming river and stream temperatures will threaten ecosystems and species, including salmon populations.[13] Salmon play a particularly important role in the diet, culture, religion, and economy of Native Americans in this region.[14]

Salmon are sensitive to water temperature at many stages of their lives. They spend much of their adult lives in the ocean, then migrate inland to spawn. Salmon need cold water to migrate and for their young to hatch and grow successfully. Warmer water can negatively affect fish, making it more difficult for them to swim upstream. It can also make fish more susceptible to disease.[15] River and stream temperatures in the Pacific Northwest are influenced by many factors, but are expected to rise as average air temperatures increase (see the U.S. and Global Temperature indicator on p. 18).[16,17,18]

The graph shows average August water temperatures at a site in the Snake River, in eastern Washington near Nez Perce tribal lands. Several species of salmon use the Snake River to migrate and spawn. Between 1960 and 2015, water temperatures have increased by 1.4°F.

Average August Temperatures in the Snake River, 1960–2015

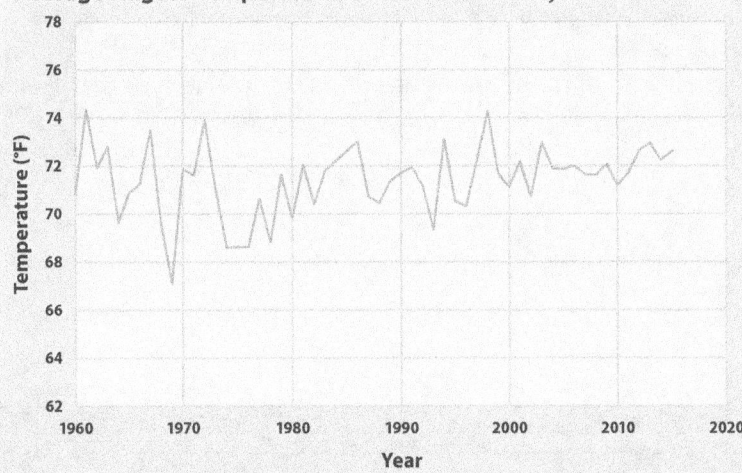

This graph shows average August water temperatures at a site along the Snake River in eastern Washington.
Data source: USGS, 2016[19]

Great Lakes Water Levels

This indicator measures water levels in the Great Lakes.

The Great Lakes, which are Lake Superior, Lake Michigan, Lake Huron, Lake Erie, and Lake Ontario, form the largest group of freshwater lakes on Earth. These lakes support a variety of ecosystems and play a vital role in the economy of the eight neighboring states and the Canadian province of Ontario, providing drinking water, shipping lanes, fisheries, recreational opportunities, and more. Water level (the height of the lake surface above sea level) is influenced by factors like precipitation, snowmelt runoff, drought, evaporation rates, and people withdrawing water for multiple uses. Warmer water, reduced ice cover, and increased evaporation resulting from climate change could affect water levels, "lake-effect" precipitation, shipping, infrastructure, and ecosystems.

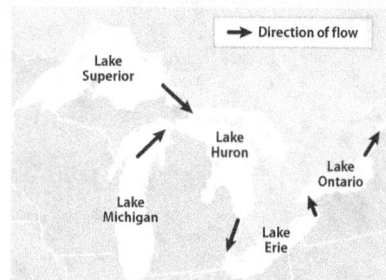

ABOUT THE INDICATOR

This indicator analyzes water levels in the Great Lakes. Water levels are recorded by gauges along the shore of each lake, some of which have been operated since the 1800s. These data were provided by the National Oceanic and Atmospheric Administration (NOAA) and the Canadian Hydrographic Service. Annual water level anomalies, or differences, in feet are compared with the average water levels in each lake from 1860 to 2015. Another component of this indicator available online tracks surface water temperatures in the Great Lakes based on satellite imagery analyzed by NOAA's Great Lakes Environmental Research Laboratory.

WHAT'S HAPPENING

- Water levels in the Great Lakes have fluctuated since 1860. Over the last few decades, they appear to have declined for most of the Great Lakes. The most recent levels are all within the range of historical variation, however.

Water Levels of the Great Lakes, 1860–2015

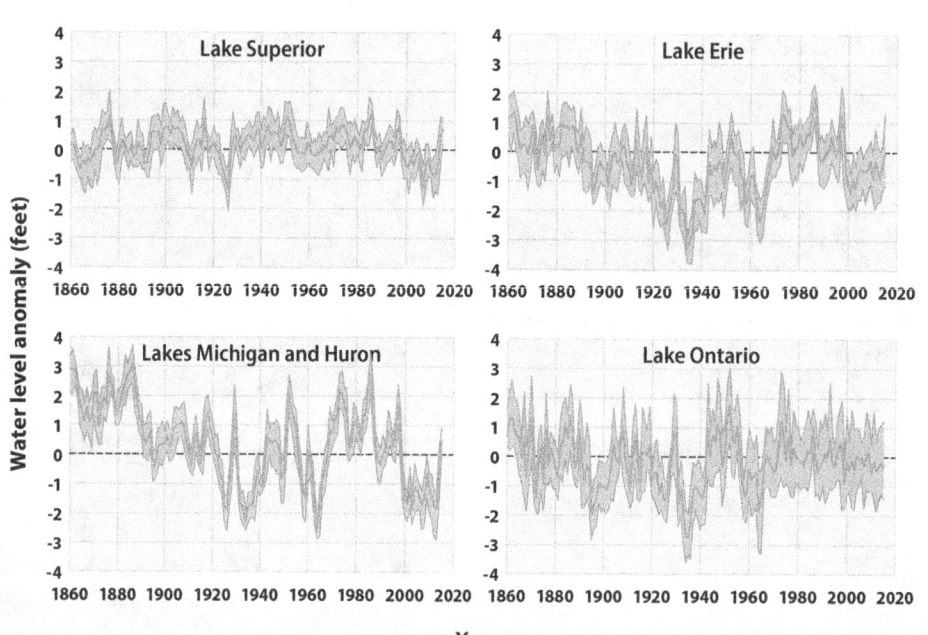

This figure displays how water levels in each of the Great Lakes have changed since 1860. For each year, the shaded band shows the range of monthly average water levels, and the line in the middle shows the annual average. The graph uses the 1981 to 2010 average as a baseline for depicting change. Choosing a different baseline period would not change the shape of the data over time. Lakes Michigan and Huron are shown together because they are connected at the same water level. Data source: NOAA, 2016[20]

Bird Wintering Ranges

This indicator examines changes in the winter ranges of North American birds.

C hanges in climate can affect ecosystems by influencing animal behavior and ranges. Birds are a particularly good indicator of these changes because the timing of certain events in their life cycles—such as migration and reproduction—is driven by cues from the environment. Changing conditions can influence the distribution of both migratory and non-migratory birds as well as the timing of important life cycle events.[21] If a change in behavior or range occurs across many types of birds, it suggests that a common external factor, such as a change in the pattern of temperature or precipitation, might be the cause. Birds are also a useful indicator because they are easy to identify and count, and thus people have kept detailed records of bird distribution and abundance for more than a century.

WHAT'S HAPPENING

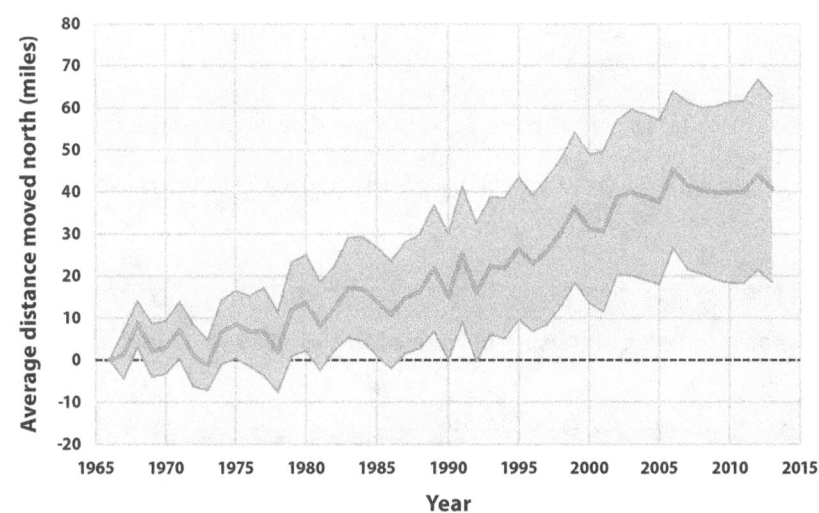

Change in Latitude of Bird Center of Abundance, 1966–2013

This figure shows annual change in latitude of bird center of abundance for 305 widespread bird species in North America from 1966 to 2013. Each winter is represented by the year in which it began (for example, winter 2013–2014 is shown as 2013). The shaded band shows the likely range of values, based on the number of measurements collected and the precision of the methods used. Data source: National Audubon Society, 2014[22]

- Among 305 widespread North American bird species, the average mid-December to early-January center of abundance moved northward by more than 40 miles between 1966 and 2013. Trends in the center of abundance moving northward can be closely related to increasing winter temperatures.[23]

- Some species have moved farther than others. A total of 48 species have moved northward by more than 200 miles. Of the 305 species studied, 186 (61 percent) have shifted their wintering grounds northward since the 1960s, while 82 (27 percent) have shifted southward. Some others have not moved at all.

ABOUT THE INDICATOR

This indicator looks collectively at the "center of abundance" of hundreds of widespread North American bird species over a 48-year period. The center of abundance is a point on the map that represents the middle of each species' distribution. If a whole population of birds were to shift generally northward, one would see the center of abundance shift northward as well. Data come from the National Audubon Society's Christmas Bird Count, which takes place every year in early winter. The Christmas Bird Count is a long-running citizen-science program in which individuals are organized by the National Audubon Society, Bird Studies Canada, local Audubon chapters, and other bird clubs to identify and count bird species at more than 2,000 locations throughout the United States and parts of Canada. At each location, observers follow a standard counting procedure to estimate the number of birds within a 15-mile diameter "count circle" over a 24-hour period. The online version of this indicator also shows how birds' wintering grounds have moved farther from the coast, which can relate to changes in winter temperatures.

Ecosystems

79

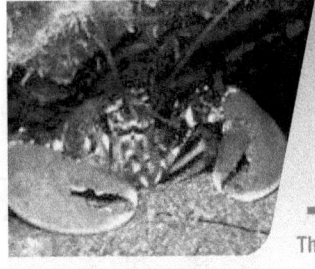

Marine Species Distribution

This indicator examines changes in the location of fish, shellfish, and other marine species along U.S. coasts.

Changes in water temperature can affect the environments where fish, shellfish, and other marine species live. As climate change causes the oceans to become warmer year-round (see the Ocean Heat and Sea Surface Temperature indicators on pp. 32 and 33), populations of some species may adapt by shifting toward cooler areas. Along U.S. coasts, this means a shift northward or to deeper waters. Marine species represent a particularly good indicator of warming oceans because they are sensitive to climate and have been studied and tracked for many years. Tracking the movement of multiple species is useful because if a change in behavior or distribution occurs across a large range of species, it is likely the result of a systematic cause rather than a species-specific one.

ABOUT THE INDICATOR

This indicator tracks marine animal species in terms of their "center of biomass," which is a point on the map that represents the center of each species' distribution by weight. If a fish population were to shift generally northward, the center of biomass would shift northward as well. Data for this indicator were collected by the National Oceanic and Atmospheric Administration's National Marine Fisheries Service and other agencies, which monitor marine species populations by conducting annual surveys at regular intervals along the coast. By recording what they catch at each location, scientists can calculate each species' center of biomass in terms of latitude, longitude, and depth. This indicator focuses on two regions that have the longest, most consistent sampling: the Northeast and the eastern Bering Sea off the coast of Alaska. The species shown in the two maps were chosen because they represent a variety of habitats and species types (a mixture of fish and shellfish), they tend to be fairly abundant, and their population trends are not unduly impacted by overfishing.

WHAT'S HAPPENING

- The average center of biomass for 105 marine fish and invertebrate species shifted northward by about 10 miles between 1982 and 2015, as shown in the chart. These species also moved an average of 20 feet deeper.

This graph shows the annual change in latitude (movement in miles) and depth of 105 marine species along the Northeast coast and in the eastern Bering Sea. Changes in the centers of biomass have been aggregated across all 105 species. Data source: NOAA and Rutgers University, 2016[24]

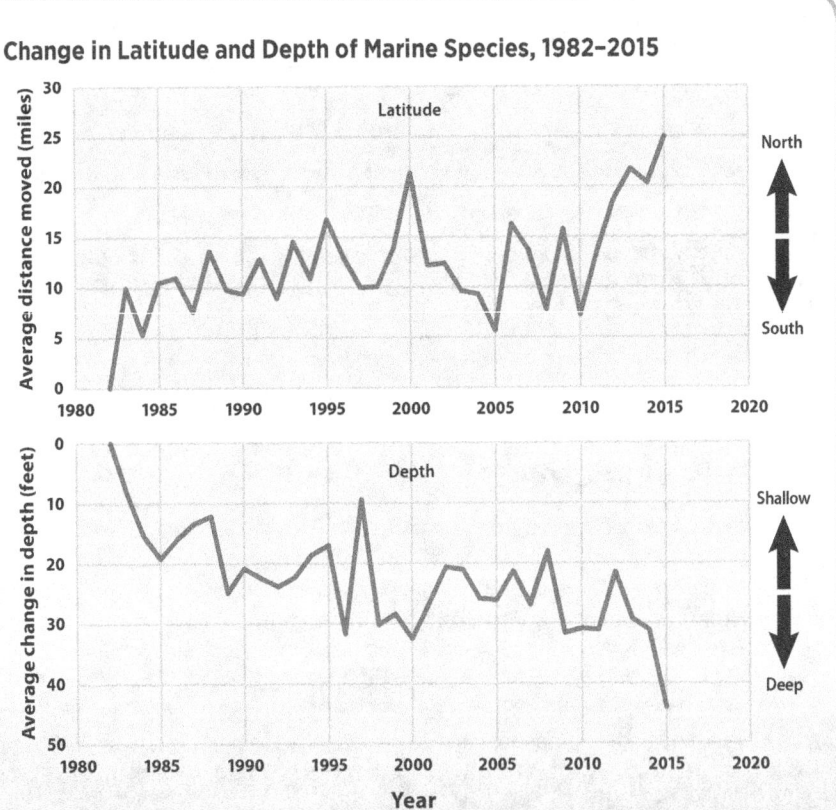

Change in Latitude and Depth of Marine Species, 1982–2015

Average Location of Three Fish and Shellfish Species in the Northeast, 1968–2015

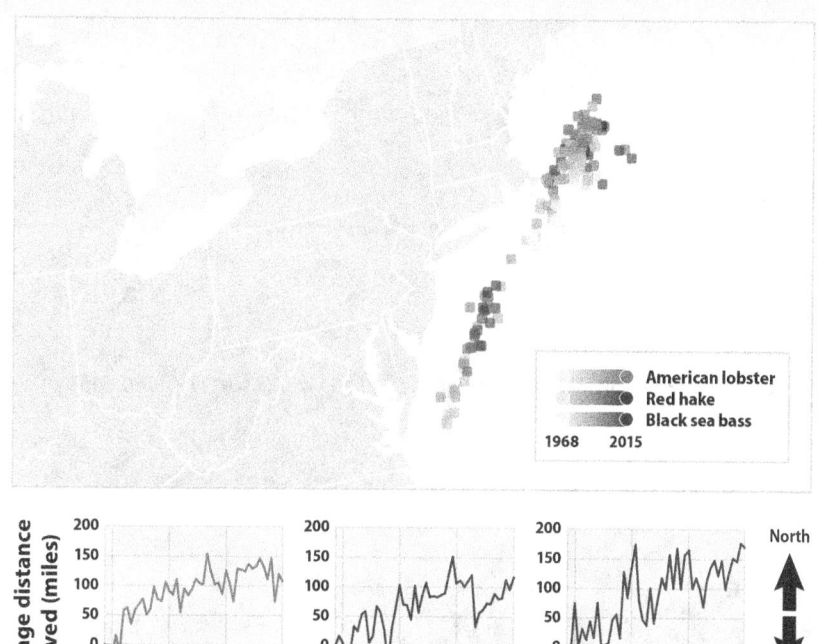

- In waters off the northeastern United States, several economically important species have shifted northward since the late 1960s, as shown in the map of the Northeast Atlantic coast. The three species shown (American lobster, red hake, and black sea bass) have moved northward by an average of 119 miles.

This map shows the annual centers of biomass for three species in the northeastern United States from 1968 to 2015. Dots are shaded from light to dark to show change over time. Visit this indicator online at: www.epa.gov/climate-indicators for an interactive version of this map. Data source: NOAA and Rutgers University, 2016[25]

Average Location of Three Fish and Shellfish Species in the Bering Sea, 1982–2015

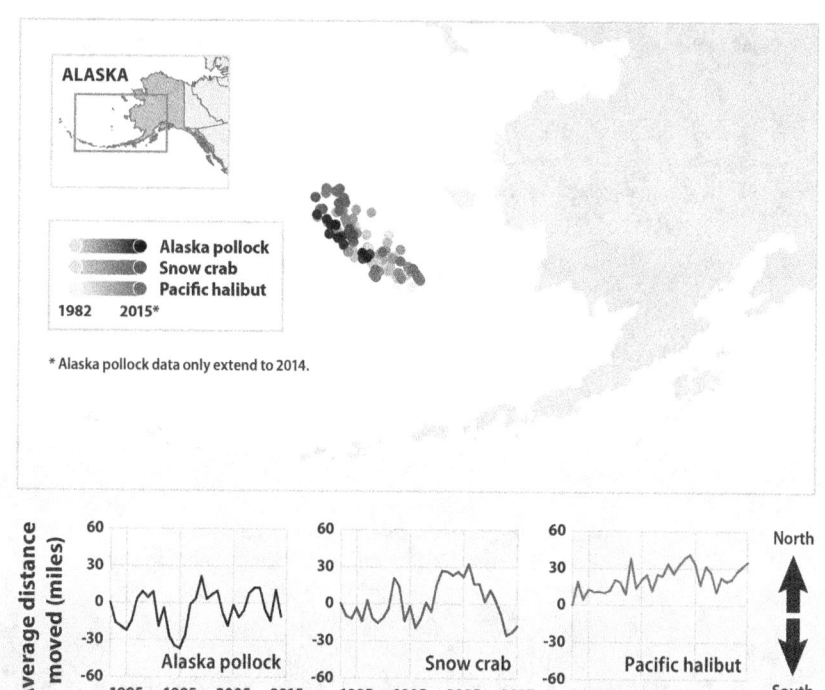

- In the Bering Sea, Alaska pollock, snow crab, and Pacific halibut have generally shifted away from the coast since the early 1980s, as shown in the map of the Bering Sea. These species have also moved northward by an average of 14 miles.

This map shows the annual centers of biomass for three species in the eastern Bering Sea from 1982 to 2015. Dots are shaded from light to dark to show change over time. Visit this indicator online at: www.epa.gov/climate-indicators for an interactive version of this map. Data source: NOAA and Rutgers University, 2016[26]

Ecosystems

81

Leaf and Bloom Dates

This indicator examines the timing of leaf growth and flower blooms for two widely distributed plants in the United States.

Phenology is the study of important seasonal events and their timing, such as flower blooms and animal migration. Phenological events are influenced by a combination of environmental factors, including temperature, light, rainfall, and humidity. Because of their close connection with climate, the timing of phenological events can be used as an indicator of the sensitivity of ecological processes to climate change. Two particularly useful indicators are the first leaf dates and the first bloom dates of lilacs and honeysuckles in the spring. Scientists have high confidence that the earlier arrival of spring events is linked to recent warming trends in global climate.[27]

WHAT'S HAPPENING

- First leaf and bloom dates in lilacs and honeysuckles in the contiguous 48 states show a great deal of year-to-year variability, which makes it difficult to determine whether a statistically meaningful change has taken place. Earlier dates appear more prevalent, however, in the last few decades.

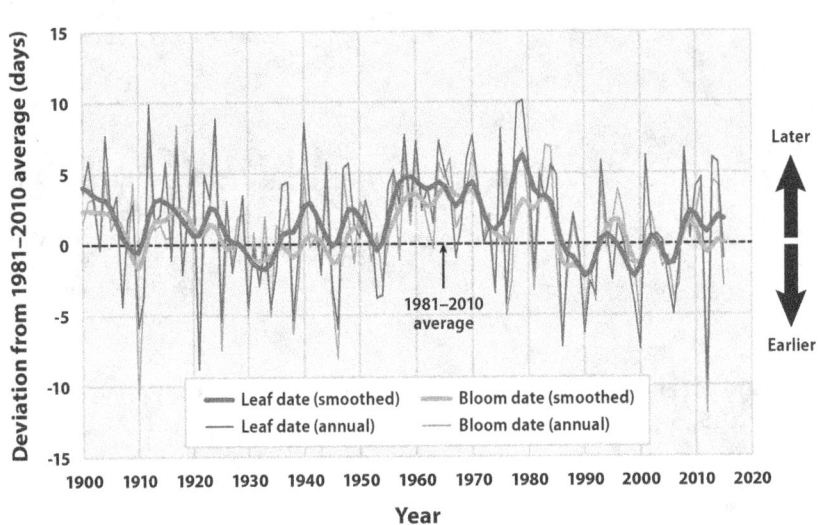

First Leaf and Bloom Dates in the Contiguous 48 States, 1900–2015

This figure shows modeled trends in lilac and honeysuckle first leaf dates and first bloom dates across the contiguous 48 states, using the 1981 to 2010 average as a baseline. Positive values indicate that leaf growth and blooming began later in the year, and negative values indicate that leafing and blooming occurred earlier. The thicker lines were smoothed using a nine-year weighted average. Choosing a different long-term average for comparison would not change the shape of the data over time. Data source: Schwartz, 2016[28]

ABOUT THE INDICATOR

This indicator shows trends in the timing of first leaf dates and first bloom dates in lilacs and honeysuckles across the contiguous 48 states. It is originally based on observations collected by the USA National Phenology Network, which collects ground observations from a network of federal agencies, field stations, educational institutions, and citizens who have been trained to log observations of leaf and bloom dates. Because many of the observation records in the United States are less than 40 years long or contain gaps, computer models have been used to provide a more complete understanding of long-term trends nationwide. These models use temperature data from thousands of weather stations, and were developed and tested based on observed relationships between leaf and bloom dates and daily temperatures. The online version of this indicator also presents maps that show how first leaf and bloom dates have changed across the country.

COMMUNITY CONNECTION: CHERRY BLOSSOM BLOOM DATES IN WASHINGTON, D.C.

In Washington, D.C., the arrival of spring brings a splash of color as the city's iconic cherry trees burst into bloom. The National Cherry Blossom Festival is planned to coincide with the peak bloom of the cherry trees and draws more than 1.5 million visitors to the area every year. The peak bloom date for the most common type of cherry tree around Washington's Tidal Basin—the Yoshino variety—has been carefully estimated and recorded since 1921 by the National Park Service. The peak bloom date is defined as the day when 70 percent of the blossoms are in full bloom. Based on the entire 96 years of data, Washington's blossoms reach their peak on April 4 in an average year. The peak bloom date has shifted earlier by approximately five days since 1921. While the length of the National Cherry Blossom Festival has continued to expand, the Yoshino cherry trees have bloomed near the beginning of the festival in recent years. During some years, the festival missed the peak bloom date entirely.

Peak Bloom Date for Cherry Trees Around Washington, D.C.'s Tidal Basin, 1921–2016

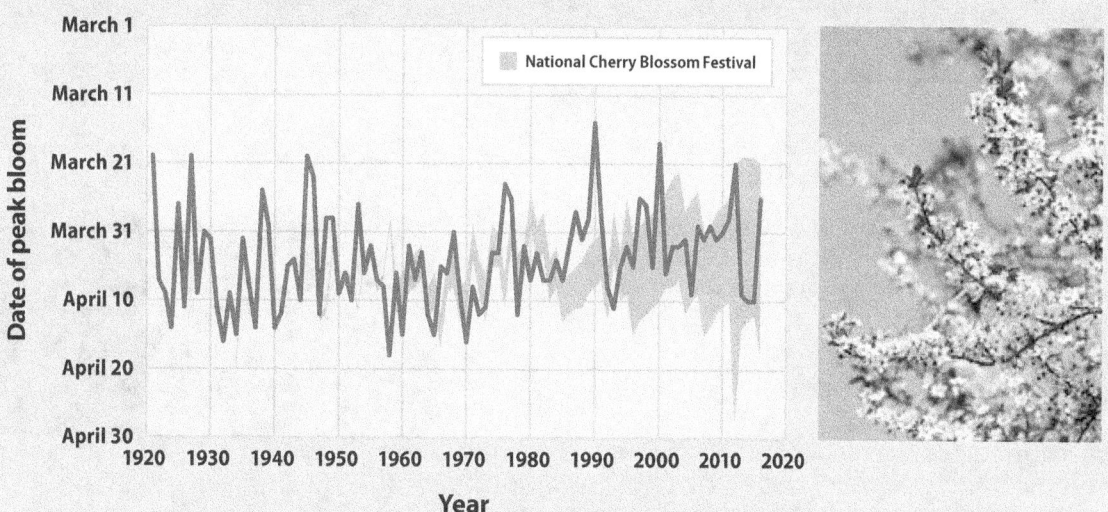

This figure shows the peak bloom date each year for the main type of cherry tree around the Tidal Basin in Washington, D.C. The peak bloom date occurs when 70 percent of the blossoms are in full bloom. The shaded band shows the timing of the annual National Cherry Blossom Festival. The festival began in 1934 but was not held during World War II. Data sources: National Cherry Blossom Festival, 2016;[29] National Park Service, 2015[30]

Climate Change Resources

EPA's Climate Change website (www.epa.gov/climate-change) provides a good starting point for further exploration of this topic. At this site, you can:

- View the latest information about EPA's climate change indicators (www.epa.gov/climate-indicators) and download figures as well as accompanying technical documentation.

- Learn more about greenhouse gases and the science of climate change, discover the potential impacts of climate change on human health and ecosystems, read about how people can adapt to changes, and get up-to-date news.

- Read about greenhouse gas emissions, look through EPA's greenhouse gas inventories, and explore EPA's Greenhouse Gas Data Publication Tool.

- Learn about EPA's regulatory initiatives and partnership programs.

- Explore U.S. climate policy and climate economics.

- Search EPA's database of frequently asked questions about climate change and ask your own questions.

- Explore a glossary of terms related to climate change, including many terms that appear in this report.

- Find out what you can do at home, on the road, at work, and at school to help reduce greenhouse gas emissions.

- Learn how you, your family, and your community can respond to and stay healthy in a changing climate.

- Find resources for educators and students.

Many other government and nongovernment websites also provide information about climate change. Here are some examples:

- The Intergovernmental Panel on Climate Change (IPCC) is the international authority on climate change science. The IPCC website (www.ipcc.ch/index.htm) summarizes the current state of scientific knowledge about climate change.

- The U.S. Global Change Research Program (www.globalchange.gov) is a multi-agency effort focused on improving our understanding of the science of climate change and its potential impacts on the United States through reports such as the *National Climate Assessment* and *The Impacts of Climate Change on Human Health in the United States: A Scientific Assessment* (https://health2016.globalchange.gov).

- The National Academy of Sciences (http://nas-sites.org/americasclimatechoices) has developed many independent scientific reports on the causes of climate change, its impacts, and potential solutions. The National Academy's Koshland Science Museum (https://koshland-science-museum.org) provides an interactive online Earth Lab where people can learn more about these topics.

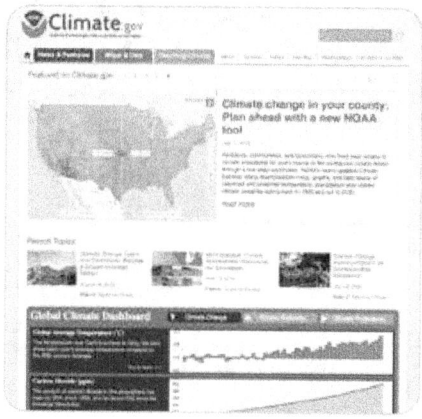

- The National Oceanic and Atmospheric Administration (NOAA) is charged with helping society understand, plan for, and respond to climate variability and change. Find out more about NOAA's climate indicators and other activities at: www.climate.gov.

- NOAA's National Centers for Environmental Information website (www.ncei.noaa.gov) provides access to data that demonstrate the effects of climate change on weather, climate, and the oceans.

- The Centers for Disease Control and Prevention (CDC) provides extensive information about the relationship between climate change and public health at: www.cdc.gov/climateandhealth/default.htm.

- The U.S. Geological Survey's Climate and Land Use Change website (www.usgs.gov/science/mission-areas/climate-and-land-use-change) looks at the relationships between natural processes on the surface of the earth, ecological systems, and human activities.

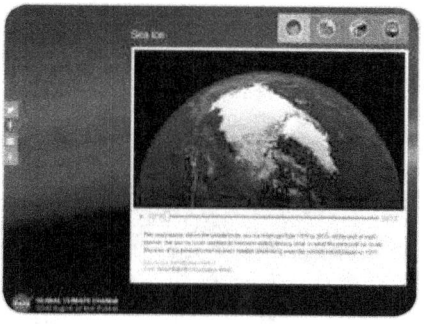

- The National Aeronautics and Space Administration (NASA) maintains its own set of climate change indicators (http://climate.nasa.gov). Another NASA site (http://earthobservatory.nasa.gov/Features/EnergyBalance/page1.php) discusses the Earth's energy budget and how it relates to greenhouse gas emissions and climate change.

- The National Snow and Ice Data Center's website (http://nsidc.org/cryosphere) provides more information about ice and snow and how they influence and are influenced by climate change.

- The Woods Hole Oceanographic Institution's website (www.whoi.edu/main/topic/climate-ocean) explains how climate change affects the oceans and how scientists measure these effects.

For more indicators of environmental condition and human health, visit EPA's Report on the Environment (www.epa.gov/roe). This resource presents a wide range of indicators of national conditions and trends in air, water, land, human exposure and health, and ecological systems.

Endnotes

INTRODUCTION

1. Melillo, J.M., T.C. Richmond, and G.W. Yohe (eds.). 2014. Climate change impacts in the United States: The third National Climate Assessment. U.S. Global Change Research Program. http://nca2014.globalchange.gov.

2. Crimmins, A., J. Balbus, J.L. Gamble, C.B. Beard, J.E. Bell, D. Dodgen, R.J. Eisen, N. Fann, M.D. Hawkins, S.C. Herring, L. Jantarasami, D.M. Mills, S. Saha, M.C. Sarofim, J. Trtanj, and L. Ziska (eds). 2016. The impacts of climate change on human health in the United States: A scientific assessment. U.S. Global Change Research Program. https://health2016.globalchange.gov.

UNDERSTANDING GREENHOUSE GASES

1. IPCC (Intergovernmental Panel on Climate Change). 2013. Climate change 2013: The physical science basis. Working Group I contribution to the IPCC Fifth Assessment Report. Cambridge, United Kingdom: Cambridge University Press. www.ipcc.ch/report/ar5/wg1.

SUMMARY OF KEY POINTS

1. IPCC (Intergovernmental Panel on Climate Change). 2013. Climate change 2013: The physical science basis. Working Group I contribution to the IPCC Fifth Assessment Report. Cambridge, United Kingdom: Cambridge University Press. www.ipcc.ch/report/ar5/wg1.

GREENHOUSE GASES

1. IPCC (Intergovernmental Panel on Climate Change). 2013. Climate change 2013: The physical science basis. Working Group I contribution to the IPCC Fifth Assessment Report. Cambridge, United Kingdom: Cambridge University Press. www.ipcc.ch/report/ar5/wg1.

2. U.S. EPA (U.S. Environmental Protection Agency). 2016. Inventory of U.S. greenhouse gas emissions and sinks: 1990–2014. EPA 430-R-16-002. www3.epa.gov/climatechange/ghgemissions/usinventoryreport.html.

3. U.S. EPA (U.S. Environmental Protection Agency). 2016. Inventory of U.S. greenhouse gas emissions and sinks: 1990–2014. EPA 430-R-16-002. www3.epa.gov/climatechange/ghgemissions/usinventoryreport.html.

4. WRI (World Resources Institute). 2014. Climate Analysis Indicators Tool (CAIT) 2.0: WRI's climate data explorer. Accessed May 2014. http://cait.wri.org.

5. FAO (Food and Agriculture Organization). 2014. FAOSTAT: Emissions—land use. Accessed May 2014. http://faostat3.fao.org/faostat-gateway/go/to/download/G2/*/E.

6. *EPICA Dome C and Vostok Station, Antarctica: approximately 796,562 BCE to 1813 CE*
    Lüthi, D., M. Le Floch, B. Bereiter, T. Blunier, J.-M. Barnola, U. Siegenthaler, D. Raynaud, J. Jouzel, H. Fischer, K. Kawamura, and T.F. Stocker. 2008. High-resolution carbon dioxide concentration record 650,000–800,000 years before present. Nature 453:379–382. www.ncdc.noaa.gov/paleo/pubs/luethi2008/luethi2008.html.

    Law Dome, Antarctica, 75-year smoothed: approximately 1010 CE to 1975 CE
    Etheridge, D.M., L.P. Steele, R.L. Langenfelds, R.J. Francey, J.-M. Barnola, and V.I. Morgan. 1998. Historical CO_2 records from the Law Dome DE08, DE08-2, and DSS ice cores. In: Trends: A compendium of data on global change. Oak Ridge, TN: U.S. Department of Energy. Accessed September 14, 2005. http://cdiac.ornl.gov/trends/co2/lawdome.html.

    Siple Station, Antarctica: approximately 1744 CE to 1953 CE
    Neftel, A., H. Friedli, E. Moor, H. Lötscher, H. Oeschger, U. Siegenthaler, and B. Stauffer. 1994. Historical carbon dioxide record from the Siple Station ice core. In: Trends: A compendium of data on global change. Oak Ridge, TN: U.S. Department of Energy. Accessed September 14, 2005. http://cdiac.ornl.gov/trends/co2/siple.html.

Mauna Loa, Hawaii: 1959 CE to 2015 CE
NOAA (National Oceanic and Atmospheric Administration). 2016. Annual mean carbon dioxide concentrations for Mauna Loa, Hawaii. Accessed April 14, 2016. ftp://ftp.cmdl.noaa.gov/products/trends/co2/co2_annmean_mlo.txt.

Barrow, Alaska: 1974 CE to 2014 CE
Cape Matatula, American Samoa: 1976 CE to 2014 CE
South Pole, Antarctica: 1976 CE to 2014 CE
NOAA (National Oceanic and Atmospheric Administration). 2016. Monthly mean carbon dioxide concentrations for Barrow, Alaska; Cape Matatula, American Samoa; and the South Pole. Accessed April 14, 2016. ftp://ftp.cmdl.noaa.gov/data/trace_gases/co2/in-situ/surface.

Cape Grim, Australia: 1992 CE to 2006 CE
Shetland Islands, Scotland: 1993 CE to 2002 CE
Steele, L.P., P.B. Krummel, and R.L. Langenfelds. 2007. Atmospheric CO_2 concentrations (ppmv) derived from flask air samples collected at Cape Grim, Australia, and Shetland Islands, Scotland. Commonwealth Scientific and Industrial Research Organisation. Accessed January 20, 2009. http://cdiac.esd.ornl.gov/ftp/trends/co2/csiro.

Lampedusa Island, Italy: 1993 CE to 2000 CE
Chamard, P., L. Ciattaglia, A. di Sarra, and F. Monteleone. 2001. Atmospheric carbon dioxide record from flask measurements at Lampedusa Island. In: Trends: A compendium of data on global change. Oak Ridge, TN: U.S. Department of Energy. Accessed September 14, 2005. http://cdiac.ornl.gov/trends/co2/lampis.html.

7. IPCC (Intergovernmental Panel on Climate Change). 2013. Climate change 2013: The physical science basis. Working Group I contribution to the IPCC Fifth Assessment Report. Cambridge, United Kingdom: Cambridge University Press. www.ipcc.ch/report/ar5/wg1.

8. NOAA (National Oceanic and Atmospheric Administration). 2016. The NOAA Annual Greenhouse Gas Index. Accessed June 2016. www.esrl.noaa.gov/gmd/aggi.

WEATHER AND CLIMATE

1. NOAA (National Oceanic and Atmospheric Administration). 2016. National Centers for Environmental Information. Accessed February 2016. www.ncei.noaa.gov.

2. NOAA (National Oceanic and Atmospheric Administration). 2016. National Centers for Environmental Information. Accessed February 2016. www.ncei.noaa.gov.

3. Melillo, J.M., T.C. Richmond, and G.W. Yohe (eds.). 2014. Climate change impacts in the United States: The third National Climate Assessment. U.S. Global Change Research Program. http://nca2014.globalchange.gov.

4. National Research Council. 2011. Climate stabilization targets: Emissions, concentrations, and impacts over decades to millennia. Washington, DC: National Academies Press.

5. NOAA (National Oceanic and Atmospheric Administration). 2015. U.S. Climate Extremes Index. Accessed December 2015. www.ncdc.noaa.gov/extremes/cei.

6. Sarofim, M.C., S. Saha, M.D. Hawkins, D.M. Mills, J. Hess, R. Horton, P. Kinney, J. Schwartz, and A. St. Juliana. 2016. Chapter 2: Temperature-related death and illness. The impacts of climate change on human health in the United States: A scientific assessment. U.S. Global Change Research Program. https://health2016.globalchange.gov.

7. NOAA (National Oceanic and Atmospheric Administration). 2016. U.S. Climate Extremes Index. Accessed May 2016. www.ncdc.noaa.gov/extremes/cei.

8. Blunden, J., and D.S. Arndt (eds.). 2016. State of the climate in 2015. B. Am. Meteorol. Soc. 97(8):S1–S275.

9. NOAA (National Oceanic and Atmospheric Administration). 2016. National Centers for Environmental Information. Accessed February 2016. www.ncei.noaa.gov.

10. Melillo, J.M., T.C. Richmond, and G.W. Yohe (eds.). 2014. Climate change impacts in the United States: The third National Climate Assessment. U.S. Global Change Research Program. http://nca2014.globalchange.gov.

11. NOAA (National Oceanic and Atmospheric Administration). 2016. U.S. Climate Extremes Index. Accessed January 2016. www.ncdc.noaa.gov/extremes/cei.

12. Bell, J.E., S.C. Herring, L. Jantarasami, C. Adrianopoli, K. Benedict, K. Conlon, V. Escobar, J. Hess, J. Luvall, C.P. Garcia-Pando, D. Quattrochi, J. Runkle, and C.J. Schreck, III. 2016. Chapter 4: Impacts of extreme events on human health. The impacts of climate change on human health in the United States: A scientific assessment. U.S. Global Change Research Program. https://health2016.globalchange.gov.

13. Melillo, J.M., T.C. Richmond, and G.W. Yohe (eds.). 2014. Climate change impacts in the United States: The third National Climate Assessment. U.S. Global Change Research Program. http://nca2014.globalchange.gov.

14. Melillo, J.M., T.C. Richmond, and G.W. Yohe (eds.). 2014. Climate change impacts in the United States: The third National Climate Assessment. U.S. Global Change Research Program. http://nca2014.globalchange.gov.

15. IPCC (Intergovernmental Panel on Climate Change). 2013. Climate change 2013: The physical science basis. Working Group I contribution to the IPCC Fifth Assessment Report. Cambridge, United Kingdom: Cambridge University Press. www.ipcc.ch/report/ar5/wg1.

16. Emanuel, K.A. 2016 update to data originally published in: Emanuel, K.A. 2007. Environmental factors affecting tropical cyclone power dissipation. J. Climate 20(22):5497–5509.

17. IPCC (Intergovernmental Panel on Climate Change). 2012. Managing the risks of extreme events and disasters to advance climate change adaptation. Cambridge, United Kingdom: Cambridge University Press. http://ipcc-wg2.gov/SREX.

18. Mallakpour, I., G. Villarini. 2015. The changing nature of flooding across the central United States. Nature Climate Change 5:250–254.

19. Slater, L., and G. Villarini. 2016 update and expansion to data originally published in: Mallakpour, I., G. Villarini. 2015. The changing nature of flooding across the central United States. Nature Climate Change 5:250–254.

20. Slater, L., and G. Villarini. 2016 update and expansion to data originally published in: Mallakpour, I., G. Villarini. 2015. The changing nature of flooding across the central United States. Nature Climate Change 5:250–254.

21. Mallakpour, I., G. Villarini. 2015. The changing nature of flooding across the central United States. Nature Climate Change 5:250–254.

22. Melillo, J.M., T.C. Richmond, and G.W. Yohe (eds.). 2014. Climate change impacts in the United States: The third National Climate Assessment. U.S. Global Change Research Program. http://nca2014.globalchange.gov.

23. Dodgen, D., D. Donato, N. Kelley, A. La Greca, J. Morganstein, J. Reser, J. Ruzek, S. Schweitzer, M.M. Shimamoto, K. Thigpen Tart, and R. Ursano. 2016. Chapter 8: Mental health and well-being. Impacts of extreme events on human health. The impacts of climate change on human health in the United States: A scientific assessment. U.S. Global Change Research Program. https://health2016.globalchange.gov.

24. Heim, R.R. 2002. A review of twentieth-century drought indices used in the United States. B. Am. Meteorol. Soc. 83(8):1149–1165.

25. NOAA (National Oceanic and Atmospheric Administration). 2016. National Centers for Environmental Information. Accessed January 2016. www7.ncdc.noaa.gov/CDO/CDODivisionalSelect.jsp.

26. Gamble, J.L., J. Balbus, M. Berger, K. Bouye, V. Campbell, K. Chief, K. Conlon, A. Crimmins, B. Flanagan, C. Gonzalez-Maddux, E. Hallisey, S. Hutchins, L. Jantarasami, S. Khoury, M. Kiefer, J. Kolling, K. Lynn, A. Manangan, M. McDonald, R. Morello-Frosch, M.H. Redsteer, P. Sheffield, K. Thigpen Tart, J. Watson, K.P. Whyte, and A.F. Wolkin. 2016. Chapter 9: Populations of concern. The impacts of climate change on human health in the United States: A scientific assessment. U.S. Global Change Research Program. https://health2016.globalchange.gov.

27. National Drought Mitigation Center. 2016. Maps and data. Accessed January 2016. http://droughtmonitor.unl.edu/MapsAndData.aspx.

28. NOAA (National Oceanic and Atmospheric Administration). 2013. State of the climate: Drought: December 2012. Accessed July 2013. www.ncdc.noaa.gov/sotc/drought/2012/12.

29. NOAA (National Oceanic and Atmospheric Administration). 2016. National Centers for Environmental Information. Accessed January 2016. www.ncdc.noaa.gov.

OCEANS

1. IPCC (Intergovernmental Panel on Climate Change). 2013. Climate change 2013: The physical science basis. Working Group I contribution to the IPCC Fifth Assessment Report. Cambridge, United Kingdom: Cambridge University Press. www.ipcc.ch/report/ar5/wg1.

2. Levitus, S., J.I. Antonov, T.P. Boyer, O.K. Baranova, H.E. Garcia, R.A. Locarnini, A.V. Mishonov, J.R. Reagan, D. Seidov, E.S. Yarosh, and M.M. Zweng. 2012. World ocean heat content and thermosteric sea level change (0–2000 m), 1955–2010. Geophys. Res. Lett. 39:L10603.

3. Levitus, S., J.I. Antonov, T.P. Boyer, O.K. Baranova, H.E. Garcia, R.A. Locarnini, A.V. Mishonov, J.R. Reagan, D. Seidov, E.S. Yarosh, and M.M. Zweng. 2012. World ocean heat content and thermosteric sea level change (0–2000 m), 1955–2010. Geophys. Res. Lett. 39:L10603.

4. Based on a total global energy supply of 13,541 million tons of oil equivalents in the year 2013, which equates to 5.7×10^{20} joules. Source: IEA (International Energy Agency). 2015. Key world energy statistics. www.iea.org/publications/freepublications/publication/KeyWorld_Statistics_2015.pdf.

5. CSIRO (Commonwealth Scientific and Industrial Research Organisation). 2016 update to data originally published in: Domingues, C.M., J.A. Church, N.J. White, P.J. Gleckler, S.E. Wijffels, P.M. Barker, and J.R. Dunn. 2008. Improved estimates of upper-ocean warming and multi-decadal sea-level rise. Nature 453:1090–1094. www.cmar.csiro.au/sealevel/thermal_expansion_ocean_heat_timeseries.html.

6. MRI/JMA (Meteorological Research Institute/Japan Meteorological Agency). 2016 update to data originally published in: Ishii, M., and M. Kimoto. 2009. Reevaluation of historical ocean heat content variations with time-varying XBT and MBT depth bias corrections. J. Oceanogr. 65:287–299.

7. NOAA (National Oceanic and Atmospheric Administration). 2016. Global ocean heat and salt content. Accessed May 2016. www.nodc.noaa.gov/OC5/3M_HEAT_CONTENT.

8. For example, see: Ostrander, G.K., K.M. Armstrong, E.T. Knobbe, D. Gerace, and E.P. Scully. 2000. Rapid transition in the structure of a coral reef community: The effects of coral bleaching and physical disturbance. P. Natl. Acad. Sci. USA. 97(10):5297–5302.

9. Pratchett, M.S., S.K. Wilson, M.L. Berumen, and M.I. McCormick. 2004. Sublethal effects of coral bleaching on an obligate coral feeding butterflyfish. Coral Reefs 23(3):352–356.

10. IPCC (Intergovernmental Panel on Climate Change). 2013. Climate change 2013: The physical science basis. Working Group I contribution to the IPCC Fifth Assessment Report. Cambridge, United Kingdom: Cambridge University Press. www.ipcc.ch/report/ar5/wg1.

11. IPCC (Intergovernmental Panel on Climate Change). 2013. Climate change 2013: The physical science basis. Working Group I contribution to the IPCC Fifth Assessment Report. Cambridge, United Kingdom: Cambridge University Press. www.ipcc.ch/report/ar5/wg1.

12. NOAA (National Oceanic and Atmospheric Administration). 2016. Extended reconstructed sea surface temperature (ERSST.v4). National Centers for Environmental Information. Accessed March 2016. www.ncdc.noaa.gov/data-access/marine-ocean-data/extended-reconstructed-sea-surface-temperature-ersst.

13. Trtanj, J., L. Jantarasami, J. Brunkard, T. Collier, J. Jacobs, E. Lipp, S. McLellan, S. Moore, H. Paerl, J. Ravenscroft, M. Sengco, and J. Thurston. 2016. Chapter 6: Climate impacts on water-related illness. The impacts of climate change on human health in the United States: A scientific assessment. U.S. Global Change Research Program. https://health2016.globalchange.gov.

14. CSIRO (Commonwealth Scientific and Industrial Research Organisation). 2015 update to data originally published in: Church, J.A., and N.J. White. 2011. Sea-level rise from the late 19th to the early 21st century. Surv. Geophys. 32:585–602. www.cmar.csiro.au/sealevel/sl_data_cmar.html.

15. NOAA (National Oceanic and Atmospheric Administration). 2016. Laboratory for Satellite Altimetry: Sea level rise. Accessed June 2016. www.star.nesdis.noaa.gov/sod/lsa/SeaLevelRise/LSA_SLR_timeseries_global.php.

16. Titus, J.G., E.K. Anderson, D.R. Cahoon, S. Gill, R.E. Thieler, and J.S. Williams. 2009. Coastal sensitivity to sea-level rise: A focus on the Mid-Atlantic region. U.S. Climate Change Science Program and the Subcommittee on Global Change Research. https://downloads.globalchange.gov/sap/sap4-1/sap4-1-final-report-all.pdf.

17. NOAA (National Oceanic and Atmospheric Administration). 2016 update to data originally published in: NOAA. 2009. Sea level variations of the United States 1854–2006. NOAA Technical Report NOS CO-OPS 053. www.tidesandcurrents.noaa.gov/publications/Tech_rpt_53.pdf.

18. NOAA (National Oceanic and Atmospheric Administration). 2013. Coastal Change Analysis Program. Accessed December 2013. https://coast.noaa.gov/dataregistry/search/collection/info/ccapregional.

19. Melillo, J.M., T.C. Richmond, and G.W. Yohe (eds.). 2014. Climate change impacts in the United States: The third National Climate Assessment. U.S. Global Change Research Program. http://nca2014.globalchange.gov.

20. NOAA (National Oceanic and Atmospheric Administration). 2016 update to data originally published in: NOAA. 2014. Sea level rise and nuisance flood frequency changes around the United States. NOAA Technical Report NOS CO-OPS 073. https://tidesandcurrents.noaa.gov/publications/NOAA_Technical_Report_NOS_COOPS_073.pdf.

21. NOAA (National Oceanic and Atmospheric Administration). 2016 update to data originally published in: NOAA. 2014. Sea level rise and nuisance flood frequency changes around the United States. NOAA Technical Report NOS CO-OPS 073. https://tidesandcurrents.noaa.gov/publications/NOAA_Technical_Report_NOS_COOPS_073.pdf.

22. Trtanj, J., L. Jantarasami, J. Brunkard, T. Collier, J. Jacobs, E. Lipp, S. McLellan, S. Moore, H. Paerl, J. Ravenscroft, M. Sengco, and J. Thurston. 2016. Chapter 6: Climate impacts on water-related illness. The impacts of climate change on human health in the United States: A scientific assessment. U.S. Global Change Research Program. https://health2016.globalchange.gov.

23. Dodgen, D., D. Donato, N. Kelley, A. La Greca, J. Morganstein, J. Reser, J. Ruzek, S. Schweitzer, M.M. Shimamoto, K. Thigpen Tart, and R. Ursano. 2016. Chapter 8: Mental health and well-being. Impacts of extreme events on human health. The impacts of climate change on human health in the United States: A scientific assessment. U.S. Global Change Research Program. https://health2016.globalchange.gov.

24. Calculated from numbers in the IPCC Fifth Assessment Report. From 1750 to present: total human emissions of 545 Pg C and ocean uptake of 155 Pg C. Source: IPCC (Intergovernmental Panel on Climate Change). 2013. Climate change 2013: The physical science basis. Working Group I contribution to the IPCC Fifth Assessment Report. Cambridge, United Kingdom: Cambridge University Press. www.ipcc.ch/report/ar5/wg1.

25. Wootton, J.T., C.A. Pfister, and J.D. Forester. 2008. Dynamic patterns and ecological impacts of declining ocean pH in a high-resolution multi-year dataset. P. Natl. Acad. Sci. USA 105(48):18848–18853.

26. Bednaršek, N., G.A. Tarling, D.C.E. Bakker, S. Fielding, E.M. Jones, H.J. Venables, P. Ward, A. Kuzirian, B. Lézé, R.A. Feely, and E.J. Murphy. 2012. Extensive dissolution of live pteropods in the Southern Ocean. Nat. Geosci. 5:881–885.

27. Bates, N.R. 2016 update to data originally published in: Bates, N.R., M.H. Best, K. Neely, R. Garley, A.G. Dickson, and R.J. Johnson. 2012. Indicators of anthropogenic carbon dioxide uptake and ocean acidification in the North Atlantic Ocean. Biogeosciences 9:2509–2522.

28. González-Dávila, M. 2012 update to data originally published in: González-Dávila, M., J.M. Santana-Casiano, M.J. Rueda, and O. Llinás. 2010. The water column distribution of carbonate system variables at the ESTOC site from 1995 to 2004. Biogeosciences 7:3067–3081.

29. Dore, J. 2015 update to data originally published in: Dore, J.E., R. Lukas, D.W. Sadler, M.J. Church, and D.M. Karl. 2009. Physical and biogeochemical modulation of ocean acidification in the central North Pacific. Proc. Natl. Acad. Sci. USA 106:12235-12240.

SNOW AND ICE

1. NASA (National Aeronautics and Space Administration). 2016. NASA's Goddard Space Flight Center Scientific Visualization Studio. https://svs.gsfc.nasa.gov.

2. NSIDC (National Snow and Ice Data Center). 2016. Sea ice data and image archive. Accessed April 2016. http://nsidc.org/data/seaice_index/archives.html.

3. NSIDC (National Snow and Ice Data Center). 2015. Arctic sea ice news and analysis. October 6, 2015. http://nsidc.org/arcticseaicenews/2015/10/2015-melt-season-in-review.

4. NASA (National Aeronautics and Space Administration). 2016. Arctic sea ice melt. http://neptune.gsfc.nasa.gov/csb/index.php?section=54.

5. IPCC (Intergovernmental Panel on Climate Change). 2013. Climate change 2013: The physical science basis. Working Group I contribution to the IPCC Fifth Assessment Report. Cambridge, United Kingdom: Cambridge University Press. www.ipcc.ch/report/ar5/wg1.

6. NSIDC (National Snow and Ice Data Center). 2016. Sea ice data and image archive. Accessed July 2016. http://nsidc.org/data/seaice_index/archives.html.

7. Parkinson, C.L. 2014. Global sea ice coverage from satellite data: Annual cycle and 35-yr trends. J. Climate 27(24):9377.

8. WGMS (World Glacier Monitoring Service). 2016 update to data originally published in: WGMS. 2015. Global glacier change bulletin no. 1 (2012–2013). Zemp, M., I. Gärtner-Roer, S.U. Nussbaumer, F. Hüsler, H. Machguth, N. Mölg, F. Paul, and M. Hoelzle (eds.). ICSU (WDS)/IUGG (IACS)/UNEP/UNESCO/WMO. Zurich, Switzerland: World Glacier Monitoring Service. http://wgms.ch/downloads/WGMS_GGCB_01.pdf.

9. O'Neel, S., E. Hood, A. Arendt, and L. Sass. 2014. Assessing streamflow sensitivity to variations in glacier mass balance. Climatic Change 123(2):329–341.

10. USGS (U.S. Geological Survey). 2015. Water resources of Alaska—glacier and snow program, benchmark glaciers. www2.usgs.gov/climate_landuse/clu_rd/glacierstudies/default.asp.

11. IPCC (Intergovernmental Panel on Climate Change). 2013. Climate change 2013: The physical science basis. Working Group I contribution to the IPCC Fifth Assessment Report. Cambridge, United Kingdom: Cambridge University Press. www.ipcc.ch/report/ar5/wg1.

12. *Cobbosseecontee Lake, Damariscotta Lake, Moosehead Lake, and Sebago Lake, Maine, 1905–2008*
Hodgkins, G.A. 2010. Historical ice-out dates for 29 lakes in New England, 1807–2008. U.S. Geological Survey Open-File Report 2010-1214.

    Cobbosseecontee Lake, Damariscotta Lake, Moosehead Lake, and Sebago Lake, Maine, 2009–2015
U.S. Geological Survey. 2016. Personal communication.

    Detroit Lake, Minnesota, 2006–2015
    Lake Osakis, Minnesota, 1905–2015
Minnesota Department of Natural Resources. Accessed June 2016. www.dnr.state.mn.us/ice_out/index.html.

    Geneva Lake, Wisconsin, 2005–2015
Geneva Lake Environmental Agency Newsletters. Accessed June 2016. www.genevaonline.com/~glea/newsletters.php.

    Lake George, New York, 2004–2015
Lake George Association. Accessed June 2016. www.lakegeorgeassociation.org/who-we-are/documents/Ice-In-Ice-Out2015.pdf.

    Lake Mendota and Lake Monona, Wisconsin, 2011–2015
North Temperate Lakes Long Term Ecological Research site. Accessed June 2016. https://lter.limnology.wisc.edu/lakeinfo/ice-data?lakeid=ME and https://lter.limnology.wisc.edu/lakeinfo/ice-data?lakeid=MO.

    Mirror Lake, New York, 2007–2012
Adirondack Daily Enterprise. Accessed December 2013. www.adirondackdailyenterprise.com.

    Mirror Lake, New York, 2013–2015
Lake Placid News. Accessed June 2016. www.lakeplacidnews.com.

    Otsego Lake, New York, 2005–2014
State University of New York (SUNY) Oneonta Biological Field Station. Annual Reports. Accessed May 2015. www.oneonta.edu/academics/biofld/PUBS/ANNUAL/2013/29-Otsego-Ice-History-2014.pdf.

    Shell Lake, Wisconsin, 2005–2015
Washburn County Clerk. 2016. Personal communication.

    All other data
NSIDC (National Snow and Ice Data Center). 2014. Global lake and river ice phenology. Last updated January 2014. http://nsidc.org/data/lake_river_ice.

13. Nenana Ice Classic. 2016. Accessed May 2016. www.nenanaakiceclassic.com.

14. Yukon River Breakup. 2016. Accessed May 2016. www.yukonriverbreakup.com.

15. Kunkel, K.E., M. Palecki, L. Ensor, K.G. Hubbard, D. Robinson, K. Redmond, and D. Easterling. 2009. Trends in twentieth-century U.S. snowfall using a quality-controlled dataset. J. Atmos. Ocean. Tech. 26:33–44.

16. NOAA (National Oceanic and Atmospheric Administration). 2016. National Centers for Environmental Information. Accessed June 2016. www.ncdc.noaa.gov.

17. Kunkel, K.E., M. Palecki, L. Ensor, K.G. Hubbard, D. Robinson, K. Redmond, and D. Easterling. 2009. Trends in twentieth-century U.S. snowfall using a quality-controlled dataset. J. Atmos. Ocean. Tech. 26:33–44.

18. Rutgers University Global Snow Lab. 2016. Area of extent data: North America (no Greenland). Accessed January 2016. http://climate.rutgers.edu/snowcover.

19. NOAA (National Oceanic and Atmospheric Administration). 2015. Snow cover maps. Accessed November 2015. ftp://eclipse.ncdc.noaa.gov/pub/cdr/snowcover.

20. Mote, P.W., and D. Sharp. 2016 update to data originally published in: Mote, P.W., A.F. Hamlet, M.P. Clark, and D.P. Lettenmaier. 2005. Declining mountain snowpack in Western North America. B. Am. Meteorol. Soc. 86(1):39–49.

HEALTH AND SOCIETY

1. Hansen, J., M. Sato, and R. Ruedy. 2012. Perception of climate change. Proc Natl Acad Sci USA 109(37):E2415–E2423.

2. Melillo, J.M., T.C. Richmond, and G.W. Yohe (eds.). 2014. Climate change impacts in the United States: The third National Climate Assessment. U.S. Global Change Research Program. http://nca2014.globalchange.gov.

3. IPCC (Intergovernmental Panel on Climate Change). 2014. Climate change 2014: Impacts, adaptation, and vulnerability. Working Group II contribution to the IPCC Fifth Assessment Report. Cambridge, United Kingdom: Cambridge University Press. www.ipcc.ch/report/ar5/wg2.

4. Sarofim, M.C., S. Saha, M.D. Hawkins, D.M. Mills, J. Hess, R. Horton, P. Kinney, J. Schwartz, and A. St. Juliana. 2016. Chapter 2: Temperature-related death and illness. The impacts of climate change on human health in the United States: A scientific assessment. U.S. Global Change Research Program. https://health2016.globalchange.gov.

5. CDC (U.S. Centers for Disease Control and Prevention). 2016. CDC WONDER database: Compressed mortality file, underlying cause of death. Accessed February 2016. http://wonder.cdc.gov/mortSQL.html.

6. CDC (U.S. Centers for Disease Control and Prevention). 2016. Indicator: Heat-related mortality. National Center for Health Statistics. Annual national totals provided by National Center for Environmental Health staff in June 2016. http://ephtracking.cdc.gov/showIndicatorPages.action.

7. Gamble, J.L., J. Balbus, M. Berger, K. Bouye, V. Campbell, K. Chief, K. Conlon, A. Crimmins, B. Flanagan, C. Gonzalez-Maddux, E. Hallisey, S. Hutchins, L. Jantarasami, S. Khoury, M. Kiefer, J. Kolling, K. Lynn, A. Manangan, M. McDonald, R. Morello-Frosch, M.H. Redsteer, P. Sheffield, K. Thigpen Tart, J. Watson, K.P. Whyte, and A.F. Wolkin. 2016. Chapter 9: Populations of concern. The impacts of climate change on human health in the United States: A scientific assessment. U.S. Global Change Research Program. https://health2016.globalchange.gov.

8. CDC (U.S. Centers for Disease Control and Prevention). 2016. CDC WONDER database: Multiple cause of death file. Accessed July 2016. http://wonder.cdc.gov/mcd-icd10.html.

9. Sarofim, M.C., S. Saha, M.D. Hawkins, D.M. Mills, J. Hess, R. Horton, P. Kinney, J. Schwartz, and A. St. Juliana. 2016. Chapter 2: Temperature-related death and illness. The impacts of climate change on human health in the United States: A scientific assessment. U.S. Global Change Research Program. https://health2016.globalchange.gov.

10. Melillo, J.M., T.C. Richmond, and G.W. Yohe (eds.). 2014. Climate change impacts in the United States: The third National Climate Assessment. U.S. Global Change Research Program. http://nca2014.globalchange.gov.

11. Melillo, J.M., T.C. Richmond, and G.W. Yohe (eds.). 2014. Climate change impacts in the United States: The third National Climate Assessment. U.S. Global Change Research Program. http://nca2014.globalchange.gov.

12. IPCC (Intergovernmental Panel on Climate Change). 2014. Climate change 2014: Impacts, adaptation, and vulnerability. Working Group II contribution to the IPCC Fifth Assessment Report. Cambridge, United Kingdom: Cambridge University Press. www.ipcc.ch/report/ar5/wg2.

13. Choudhary, E. and A. Vaidyanathan. 2014. Heat stress illness hospitalizations—Environmental public health tracking program, 20 states, 2001–2010. Surveillance Summaries 63(SS13):1–10. www.cdc.gov/mmwr/preview/mmwrhtml/ss6313a1.htm.

14. Choudhary, E. and A. Vaidyanathan. 2014. Heat stress illness hospitalizations—Environmental public health tracking program, 20 states, 2001–2010. Surveillance Summaries 63(SS13):1–10. www.cdc.gov/mmwr/preview/mmwrhtml/ss6313a1.htm.

15. CDC (U.S. Centers for Disease Control and Prevention). 2016. Environmental Public Health Tracking Program: Heat stress hospitalizations indicator data. Accessed March 2016. http://ephtracking.cdc.gov/showHome.action.

16. Choudhary, E. and A. Vaidyanathan. 2014. Heat stress illness hospitalizations—Environmental public health tracking program, 20 states, 2001–2010. Surveillance Summaries 63(SS13):1–10. www.cdc.gov/mmwr/preview/mmwrhtml/ss6313a1.htm.

17. Choudhary, E. and A. Vaidyanathan. 2014. Heat stress illness hospitalizations—Environmental public health tracking program, 20 states, 2001–2010. Surveillance Summaries 63(SS13):1–10. www.cdc.gov/mmwr/preview/mmwrhtml/ss6313a1.htm.

18. Choudhary, E. and A. Vaidyanathan. 2014. Heat stress illness hospitalizations—Environmental public health tracking program, 20 states, 2001–2010. Surveillance Summaries 63(SS13):1–10. www.cdc.gov/mmwr/preview/mmwrhtml/ss6313a1.htm.

19. NOAA (National Oceanic and Atmospheric Administration). 2016. National Centers for Environmental Information. Accessed January 2016. www.ncei.noaa.gov.

20. Beard, C.B., R.J. Eisen, C.M. Barker, J.F. Garofalo, M. Hahn, M. Hayden, A.J. Monaghan, N.H. Ogden, and P.J. Schramm. 2016. Chapter 5: Vectorborne diseases. The impacts of climate change on human health in the United States: A scientific assessment. U.S. Global Change Research Program. https://health2016.globalchange.gov.

21. Beard, C.B., R.J. Eisen, C.M. Barker, J.F. Garofalo, M. Hahn, M. Hayden, A.J. Monaghan, N.H. Ogden, and P.J. Schramm. 2016. Chapter 5: Vectorborne diseases. The impacts of climate change on human health in the United States: A scientific assessment. U.S. Global Change Research Program. https://health2016.globalchange.gov.

22. CDC (U.S. Centers for Disease Control and Prevention). 2015. Lyme disease data and statistics. www.cdc.gov/lyme/stats/index.html. Accessed December 2015.

23. CDC (U.S. Centers for Disease Control and Prevention). 2015. Lyme disease data and statistics. www.cdc.gov/lyme/stats/index.html. Accessed December 2015.

24. Gamble, J.L., J. Balbus, M. Berger, K. Bouye, V. Campbell, K. Chief, K. Conlon, A. Crimmins, B. Flanagan, C. Gonzalez-Maddux, E. Hallisey, S. Hutchins, L. Jantarasami, S. Khoury, M. Kiefer, J. Kolling, K. Lynn, A. Manangan, M. McDonald, R. Morello-Frosch, M.H. Redsteer, P. Sheffield, K. Thigpen Tart, J. Watson, K.P. Whyte, and A.F. Wolkin. 2016. Chapter 9: Populations of concern. The impacts of climate change on human health in the United States: A scientific assessment. U.S. Global Change Research Program. https://health2016.globalchange.gov.

25. Beard, C.B., R.J. Eisen, C.M. Barker, J.F. Garofalo, M. Hahn, M. Hayden, A.J. Monaghan, N.H. Ogden, and P.J. Schramm. 2016. Chapter 5: Vector-borne diseases. The impacts of climate change on human health in the United States: A scientific assessment. U.S. Global Change Research Program. https://health2016.globalchange.gov.

26. CDC (U.S. Centers for Disease Control and Prevention). 2016. West Nile virus symptoms and treatment. www.cdc.gov/westnile/symptoms/index.html. Accessed January 2016.

27. Beard, C.B., R.J. Eisen, C.M. Barker, J.F. Garofalo, M. Hahn, M. Hayden, A.J. Monaghan, N.H. Ogden, and P.J. Schramm. 2016. Chapter 5: Vectorborne diseases. The impacts of climate change on human health in the United States: A scientific assessment. U.S. Global Change Research Program. https://health2016.globalchange.gov.

28. Beard, C.B., R.J. Eisen, C.M. Barker, J.F. Garofalo, M. Hahn, M. Hayden, A.J. Monaghan, N.H. Ogden, and P.J. Schramm. 2016. Chapter 5: Vectorborne diseases. The impacts of climate change on human health in the United States: A scientific assessment. U.S. Global Change Research Program. https://health2016.globalchange.gov.

29. Beard, C.B., R.J. Eisen, C.M. Barker, J.F. Garofalo, M. Hahn, M. Hayden, A.J. Monaghan, N.H. Ogden, and P.J. Schramm. 2016. Chapter 5: Vectorborne diseases. The impacts of climate change on human health in the United States: A scientific assessment. U.S. Global Change Research Program. https://health2016.globalchange.gov.

30. CDC (U.S. Centers for Disease Control and Prevention). 2016. West Nile virus statistics and maps. www.cdc.gov/westnile/statsmaps/index.html. Accessed January 2016.

31. CDC (U.S. Centers for Disease Control and Prevention). 2016. West Nile virus statistics and maps. www.cdc.gov/westnile/statsmaps/index.html. Accessed January 2016.

32. Gamble, J.L., J. Balbus, M. Berger, K. Bouye, V. Campbell, K. Chief, K. Conlon, A. Crimmins, B. Flanagan, C. Gonzalez-Maddux, E. Hallisey, S. Hutchins, L. Jantarasami, S. Khoury, M. Kiefer, J. Kolling, K. Lynn, A. Manangan, M. McDonald, R. Morello-Frosch, M.H. Redsteer, P. Sheffield, K. Thigpen Tart, J. Watson, K.P. Whyte, and A.F. Wolkin. 2016. Chapter 9: Populations of concern. The impacts of climate change on human health in the United States: A scientific assessment. U.S. Global Change Research Program. https://health2016.globalchange.gov.

33. Dodgen, D., D. Donato, N. Kelley, A. La Greca, J. Morganstein, J. Reser, J. Ruzek, S. Schweitzer, M.M. Shimamoto, K. Thigpen Tart, and R. Ursano. 2016. Chapter 8: Mental health and well-being. Impacts of extreme events on human health. The impacts of climate change on human health in the United States: A scientific assessment. U.S. Global Change Research Program. https://health2016.globalchange.gov.

34. Beard, C.B., R.J. Eisen, C.M. Barker, J.F. Garofalo, M. Hahn, M. Hayden, A.J. Monaghan, N.H. Ogden, and P.J. Schramm. 2016. Chapter 5: Vectorborne diseases. The impacts of climate change on human health in the United States: A scientific assessment. U.S. Global Change Research Program. https://health2016.globalchange.gov.

35. IPCC (Intergovernmental Panel on Climate Change). 2014. Climate change 2014: Impacts, adaptation, and vulnerability. Working Group II contribution to the IPCC Fifth Assessment Report. Cambridge, United Kingdom: Cambridge University Press. www.ipcc.ch/report/ar5/wg2.

36. Kunkel, K.E. 2016 expanded analysis of data originally published in: Kunkel, K.E., D.R. Easterling, K. Hubbard, and K. Redmond. 2004. Temporal variations in frost-free season in the United States: 1895–2000. Geophys. Res. Lett. 31:L03201.

37. Kunkel, K.E. 2016 update to data originally published in: Kunkel, K.E., D.R. Easterling, K. Hubbard, and K. Redmond. 2004. Temporal variations in frost-free season in the United States: 1895–2000. Geophys. Res. Lett. 31:L03201.

38. Fann, N., T. Brennan, P. Dolwick, J.L. Gamble, V. Ilacqua, L. Kolb, C.G. Nolte, T.L. Spero, and L. Ziska. 2016. Chapter 3: Air quality impacts. The impacts of climate change on human health in the United States: A scientific assessment. U.S. Global Change Research Program. https://health2016.globalchange.gov.

39. IPCC (Intergovernmental Panel on Climate Change). 2013. Climate change 2013: The physical science basis. Working Group I contribution to the IPCC Fifth Assessment Report. Cambridge, United Kingdom: Cambridge University Press. www.ipcc.ch/report/ar5/wg1.

40. Ziska, L., K. Knowlton, C. Rogers, D. Dalan, N. Tierney, M. Elder, W. Filley, J. Shropshire, L.B. Ford, C. Hedberg, P. Fleetwood, K.T. Hovanky, T. Kavanaugh, G. Fulford, R.F. Vrtis, J.A. Patz, J. Portnoy, F. Coates, L. Bielory, and D. Frenz. 2011. Recent warming by latitude associated with increased length of ragweed pollen season in central North America. P Natl. Acad. Sci. USA 108:4248–4251.

41. Ziska, L., K. Knowlton, C. Rogers, National Allergy Bureau, Aerobiology Research Laboratories, Canada. 2016 update to data originally published in: Ziska, L., K. Knowlton, C. Rogers, D. Dalan, N. Tierney, M. Elder, W. Filley, J. Shropshire, L.B. Ford, C. Hedberg, P. Fleetwood, K.T. Hovanky, T. Kavanaugh, G. Fulford, R.F. Vrtis, J.A. Patz, J. Portnoy, F. Coates, L. Bielory, and D. Frenz. 2011. Recent warming by latitude associated with increased length of ragweed pollen season in central North America. P Natl. Acad. Sci. USA 108:4248–4251.

42. Fann, N., T. Brennan, P. Dolwick, J.L. Gamble, V. Ilacqua, L. Kolb, C.G. Nolte, T.L. Spero, and L. Ziska, 2016. Chapter 3: Air quality impacts. The impacts of climate change on human health in the United States: A scientific assessment. U.S. Global Change Research Program. https://health2016.globalchange.gov.

43. Ziska, L., K. Knowlton, C. Rogers, D. Dalan, N. Tierney, M. Elder, W. Filley, J. Shropshire, L.B. Ford, C. Hedberg, P. Fleetwood, K.T. Hovanky, T. Kavanaugh, G. Fulford, R.F. Vrtis, J.A. Patz, J. Portnoy, F. Coates, L. Bielory, and D. Frenz. 2011. Recent warming by latitude associated with increased length of ragweed pollen season in central North America. P Natl. Acad. Sci. USA 108:4248–4251.

ECOSYSTEMS

1. NIFC (National Interagency Fire Center). 2016. Total wildland fires and acres (1960–2015). Accessed March 2016. www.nifc.gov/fireInfo/fireInfo_stats_totalFires.html.

2. Short, K.C. 2015. Sources and implications of bias and uncertainty in a century of U.S. wildfire activity data. Int. J. Wildland Fire 24(7):883–891.

3. MTBS (Monitoring Trends in Burn Severity). 2016. MTBS data summaries. www.mtbs.gov/data/search.html.

4. Fann, N., T. Brennan, P. Dolwick, J.L. Gamble, V. Ilacqua, L. Kolb, C.G. Nolte, T.L. Spero, and L. Ziska, 2016. Chapter 3: Air quality impacts. The impacts of climate change on human health in the United States: A scientific assessment. U.S. Global Change Research Program. https://health2016.globalchange.gov.

5. USGS (U.S. Geological Survey). 2016. Analysis of data from the National Water Information System. Accessed May 2016.

6. USGS (U.S. Geological Survey). 2016. Analysis of data from the National Water Information System. Accessed May 2016.

7. USGS (U.S. Geological Survey). 2016. Analysis of data from the National Water Information System. Accessed May 2016.

8. USGS (U.S. Geological Survey). 2016. Analysis of data from the National Water Information System. Accessed May 2016.

9. Duan, S.W., and S.S. Kaushal. 2013. Warming increases carbon and nutrient fluxes from sediments in streams across land use. Biogeosciences 10:1193–1207.

10. Duan, S.W., and S.S. Kaushal. 2013. Warming increases carbon and nutrient fluxes from sediments in streams across land use. Biogeosciences 10:1193–1207.

11. Jastram, J.D., and K.C. Rice. 2015. Air- and stream-water-temperature trends in the Chesapeake Bay region, 1960–2014. U.S. Geological Survey Open-File Report 2015-1207. https://pubs.er.usgs.gov/publication/ofr20151207.

12. Jastram, J.D., and K.C. Rice. 2015. Air- and stream-water-temperature trends in the Chesapeake Bay region, 1960–2014. U.S. Geological Survey Open-File Report 2015-1207. https://pubs.er.usgs.gov/publication/ofr20151207.

13. Melillo, J.M., T.C. Richmond, and G.W. Yohe (eds.). 2014. Climate change impacts in the United States: The third National Climate Assessment. U.S. Global Change Research Program. http://nca2014.globalchange.gov.

14. Dittmer, K. 2013. Changing streamflow on Columbia basin tribal lands—Climate change and salmon. Climatic Change 120(3):627–641.

15. U.S. EPA (U.S. Environmental Protection Agency). 2001. Issue paper 5: Summary of technical literature examining the physiological effects of temperature on salmonids. EPA-910-D-01-005.

16. Caissie, D. 2006. The thermal regime of rivers: a review. Freshwater Biology 51:1389–1406.

17. Van Vliet, M.T.H., F. Ludwig, and P. Kabat. 2013. Global streamflow and thermal habitats of freshwater fishes under climate change. Climatic Change 121:739–754.

18. Isaak, D.J., S. Wollrab, D. Horan, and G. Chandler. 2012. Climate change effects on stream and river temperatures across the Northwest U.S. from 1980–2009 and implications for salmonid fishes. Climatic Change 113:499–524.

19. USGS (U.S. Geological Survey). 2016. Analysis of data from the National Water Information System. Accessed March 2016.

20. NOAA (National Oceanic and Atmospheric Administration). 2016. Great Lakes water level observations. Accessed April 2016. www.glerl.noaa.gov/data/dashboard/data.

21. La Sorte, F.A., and F.R. Thompson III. 2007. Poleward shifts in winter ranges of North American birds. Ecology 88(7):1803–1812.

22. National Audubon Society. 2014 update to data originally published in: National Audubon Society. 2009. Northward shifts in the abundance of North American birds in early winter: A response to warmer winter temperatures? http://web4.audubon.org/bird/bacc/techreport.html.

23. National Audubon Society. 2009. Northward shifts in the abundance of North American birds in early winter: A response to warmer winter temperatures? http://web4.audubon.org/bird/bacc/techreport.html.

24. NOAA (National Oceanic and Atmospheric Administration) and Rutgers University. 2016. OceanAdapt. http://oceanadapt.rutgers.edu.

25. NOAA (National Oceanic and Atmospheric Administration) and Rutgers University. 2016. OceanAdapt. http://oceanadapt.rutgers.edu.

26. NOAA (National Oceanic and Atmospheric Administration) and Rutgers University. 2016. OceanAdapt. http://oceanadapt.rutgers.edu.

27. IPCC (Intergovernmental Panel on Climate Change). 2014. Climate change 2014: Impacts, adaptation, and vulnerability. Working Group II contribution to the IPCC Fifth Assessment Report. Cambridge, United Kingdom: Cambridge University Press. www.ipcc.ch/report/ar5/wg2.

28. Schwartz, M.D. 2016 update to data originally published in: Schwartz, M.D., T.R. Ault, and J.L. Betancourt. 2013. Spring onset variations and trends in the continental United States: Past and regional assessment using temperature-based indices. Int. J. Climatol. 33:2917–2922.

29. National Cherry Blossom Festival. 2016. Bloom watch. Accessed April 14, 2016. www.nationalcherryblossomfestival.org/about/bloom-watch.

30. National Park Service. 2015. Bloom schedule. Accessed April 24, 2015.

Photo Credits